Kraft Pulping

A Compilation of Notes

Second Printing
Revised

Agneta Mimms, Compiler/Editor
Dr. Michael J. Kocurek, Technical Editor
Dr. Jeff A. Pyatte, Editor
Dr. Elizabeth E. Wright, Editor

TAPPI
Atlanta, GA

TAPPI PRESS

The Association assumes no liability or responsibility in connection with the use of this information or data, including, but not limited to, any liability or responsibility under patent, copyright, or trade secret laws. The user is responsible for determining that this document is the most recent edition published.

Within the context of this work, the author(s) may use as examples specific manufacturers of equipment. This does not imply that these manufacturers are the only or best sources of the equipment or that TAPPI endorses them in any way. The presentation of such material by TAPPI should not be construed as an endorsement of or suggestion for any agreed upon course of conduct or concerted action.

Second Printing, Revised

TAPPI PRESS
Technology Park/Atlanta
P.O. Box 105113
Atlanta, GA 30348-5113, U.S.A.

ISBN 0-89852-322-2 • TAPPI PRESS Item Number 01 01 R171
Printed in the United States of America

Library of Congress Cataloging-in-Publication Data

Kraft pulping: a compilation of notes/Agneta Mimms, complier/editor ... [et al.].
p. cm.
Includes bibliographical references.
ISBN 0-89852-053-3 (pbk.)
1. Sulphate pulping process I. Mimms, Agneta
TS1176.6.S9K7 1990
676' 126--dc20 89-20694
CIP

Preface

Some time ago when the editors became involved in writing and producing an interactive videodisc course on Kraft Pulping for TAPPI, we realized there were few single sources a student could turn to for definitive pulping information. While we found excellent sources for one topic or another, usually in the midst of a larger body of information, we did not find one source containing the information required to support the instructional content presented in the videodisc courses.

The first step in the development of the videodisc courses was to *collect* and *assemble* the content upon which the instructional strategies would be based. The project team set about the task of perusing the literature and isolating those sources that best presented the information either through word, chart, or graphic. The next step was to collate this information into a workable document from which the videodisc scripts would be written.

Upon completion, we realized that this compilation of notes might serve a purpose, in and of themselves, to students and others interested in the subject of Kraft Pulping. We felt the students taking the three courses in TAPPI's *Kraft Pulping: A Key to Quality Production* would certainly benefit, because it was not possible or desirable to present all needed information via video, especially charts and lengthy explanations best suited to the printed page.

This compilation of notes is not presented as an original work; although, many passages and charts were created in an effort to bring order and meaning to disparate explanations. In some places, the narrative resembles exactly what it is, a collection of notes, rather than a full discourse. Our primary concern was bringing the information together in one place. The content was drawn primarily from twenty TAPPI sources. The editors wish to document and acknowledge these sources, for without them, this compilation of notes would not have been possible. See the list of contributors for the sources used.

As Project Director, I would like to specifically acknowledge the contribution of the team members. Agneta Mimms was primarily responsible for collecting and assembling the written material from various sources. Her many hours of meticulous work are applauded and acknowledged. We especially appreciate Dr. Michael J. Kocurek, Herty Foundation, for suggesting the content notes be organized for publication and for his editing the final manuscript for technical accuracy. In addition to Dr. Kocurek, we wish to thank Dr. Peter E. Parker, Dr. Terry N. Adams, and William S. Fuller for their valuable information and assistance throughout the project. Dr. Elizabeth E. Wright served as instructional designer and worked with me to edit and finalize the manuscript for publication.

As you read these notes, I encourage you to also read the original source material for a more thorough and pointed presentation of each individual topic. The topics presented will be covered in more than one source; that is why you are encouraged to read all contributors listed as a next step in your study of Kraft pulping. After this, I recommend that you review and test your knowledge by taking the four TAPPI companion courses to this publication *Kraft Pulping: A Key to Quality Production:*

"Kraft Pulping Principles"
"Kraft Pulping Processes: Part 1"
"Kraft Pulping Processes: Part 2"
"Kraft Pulping Simulation"

These interactive videodisc courses were developed by Tratec, Inc., of Destin, Florida, by the following project team:

Jeff A. Pyatte, Ph.D.	Project Director
Elizabeth E. Wright, Ph.D.	Instructional Designer Simulation Script Development
Agneta Mimms	Research Simulation Script Development
Rick Mauney	Script Development
Kevin Wilson and Jennifer Wauson "The Production Group"	Videodisc Production
Stan Jarvis "Jarvis and Assoc."	Programming
Industrial Training Corporation	Simulation Programming

SPECIAL APPRECIATION TO:

Michael J. Kocurek, Ph.D.	Technical Review
Terry N. Adams, Ph.D.	Technical Review
Peter E. Parker, Ph.D.	Technical Review
William S. Fuller	Technical Review

September 25, 1989 Jeff A. Pyatte, Ph.D.

Contributors

Casey, J.P., Ed. *Pulp and Paper Chemistry and Chemical Technology.* Vol. 1. 3rd ed. New York: John Wiley & Sons, 1980.

Cowan, B., "Post-Digester Treatment of Sulfate Pulp," in *Pulp and Paper Manufacture.* Edited by T.M. Grace, B. Leopold, E.W. Malcolm and M.J. Kocurek. Vol. 5: *Alkaline Pulping.* 3rd ed. Atlanta and Montreal: Joint Textbook of the Paper Industry, 1989.

Cowan, B. "The Screening of Chemical Pulp," in *Pulp and Paper Manufacture.* Edited by T.M. Grace, B. Leopold, E.W. Malcolm and M.J. Kocurek. Vol. 5: *Alkaline Pulping.* 3rd ed. Atlanta and Montreal: Joint Textbook of the Paper Industry, 1989.

Crotogino, R.H., Poirier, N.A., and Trinh, D.T. "The Principles of Pulp Washing." *Tappi J.* 70(6): (June 1987):95.

Grace, T.M., Leopold, B., Malcolm, E.W., Kocurek, M.J., eds. "The Chemistry of Alkaline Pulping," in *Pulp and Paper Manufacture.* Vol. 5: *Alkaline Pulping.* 3rd ed. Atlanta and Montreal: Joint Textbook of the Paper Industry, 1989.

Hamilton, F., and Leopold, B., eds. *Pulp and Paper Manufacture.* Vol. 3: *Secondary Fibers and Non-Wood Pulping.* 3rd ed. Atlanta and Montreal: Joint Textbook of the Paper Industry, 1987.

Hatton, J.V., Ed. "Chip Quality Monograph," in *Pulp and Paper Technology Series No. 5.* Atlanta and Montreal: Joint Textbook of the Paper Industry, 1979.

Hooper, A.W., "The Screening of Chemical Pulp," in *Pulp and Paper Manufacture.* Edited by T.M. Grace, B. Leopold, E.W. Malcolm, and M.J. Kocurek. Vol. 5: *Alkaline Pulping.* 3rd ed. Atlanta and Montreal: Joint Textbook of the Paper Industry, 1989.

Ingruber, O.V., "Alkaline Digester System," in *Pulp and Paper Manufacture.* Edited by T.M. Grace, B. Leopold, E.W. Malcolm, and M.J. Kocurek. Vol. 5: *Akaline Pulping.* 3rd ed. Atlanta and Montreal: Joint Textbook of the Paper Industry, 1989.

Ingruber, O.V., Kocurek, M.J., and Wong, A., eds. *Pulp and Paper Manufacture.* Vol. 4: *Sulfite Science and Technology.* 3rd ed. Atlanta and Montreal: Joint Textbook of the Paper Industry, 1985.

Kocurek, M.J., and Stevens, C.F.B., eds. *Pulp and Paper Manufacture.* Vol. 1: *Properties of Fibrous Raw Materials and Their Preparation for Pulping.* 3rd ed. Atlanta and Montreal: Joint Textbook of the Paper Industry, 1983.

Leask, R.A., and Kocurek, M.J., eds. *Pulp and Paper Manufacture.* Vol. 2: *Mechanical Pulping.* 3rd ed. Atlanta and Montreal: Joint Textbook of the Paper Industry, 1987.

Perkins, J.K., ed. *Brown Stock Washing Using Rotary Filters.* Atlanta: Tappi Press, 1983.

Perkins, J.K. "Post Digester Treatment of Sulphate Pulp," in *Pulp and Paper Manufacture.* Edited by T.M. Grace, B. Leopold, E.W. Malcolm, and M.J. Kocurek. Vol. 5: *Alkaline Pulping.* 3rd ed. Atlanta and Montreal: Joint Textbook of the Paper Industry, 1989.

Perkins, J.K. "Brown Stock Washing," in *Pulp and Paper Manufacture.* Edited by T.M. Grace, B. Leopold, E.W. Malcolm, and M.J. Kocurek. Vol. 5: *Alkaline Pulping.* 3rd ed. Atlanta and Montreal: Joint Textbook of the Paper Industry, 1989.

Sjödin, L. and Petterson, B. "Two Case Studies on the Cold Blow Technique for Batch Kraft Pulping." *Tappi J.* 70:(2): (February, 1987): 72–76.

Sjöstrom, E. *Wood Chemistry,* New York: Academic Press, 1981.

Smook, G.A., *Handbook for Pulp and Paper Technologists.* Edited by M.J. Kocurek. Atlanta and Montreal: Joint Textbook of the Paper Industry, 1982.

Sulfatmassetillverkning, Sveriges Skogsindustriforbund, Sweden, 1986.

TAPPI 1983 Pulping Conference. Houston, Texas: October 24–26.

TAPPI 1984 Pulping Conference. San Francisco, California: November 12–14.

TAPPI 1986 Pulping Conference. Toronto, Ontario: October 26–30.

TAPPI, *Introduction to Pulping Technology,* TAPPI Home Study Course, no. 2. Atlanta: TAPPI Press, 1976.

TAPPI, *Notes from Chip Preparation and Quality Seminar.* Washington, D.C., 1987.

Wikdahl, B. "Centrifugal Cleaning," in *Pulp and Paper Manufacture.* Edited by T.M. Grace, B. Leopold, E.W. Malcolm, and M.J. Kocurek. Vol. 5: *Alkaline Pulping,* 3rd ed. Atlanta and Montreal: Joint Textbook of the Paper Industry, 1989.

Manufacturers Cited

Jones Division, Beloit Corp.
Black Clawson Co.
C–E Bauer
Kamyr

Contents

5 Kraft Pulping: Equipment / 77

Figures

Chapter 1

Chapter 2

1
Wood Chemistry

The study of wood chemistry involves the chemical composition and cellular structure of wood. Chemically, wood can be divided into four components: cellulose, hemicellulose, lignin, and extractives. The chemical composition and the properties of these components play a major role in the pulping process.

In pulping, it is also central to understand the structure of a tree stem as well as the different types of wood and bark in cross section.

Softwoods and hardwoods consist of somewhat different kinds of cells having specific functions. Not all types of cells are good for papermaking. Certain softwood and hardwood cells make ideal fibers for papermaking; some do not.

Knowledge of the structure of one wood cell, a softwood tracheid, serves to illustrate these key pulping principles.

Chemical Components of Wood

Wood contains many different chemical substances, which can be divided into four major groups:

- Cellulose
- Hemicellulose
- Lignin
- Extractives.

When pulping, one wants to retain as much of the cellulose as possible and, normally, the hemicellulose, while lignin and extractives are the components removed from the wood fibers during pulping. The chemical composition of wood varies from species to species. In general, hardwoods contain more hemicellulose than softwoods but less lignin and extractives. Figure 1.1 shows typical chemical compositions of hardwoods and softwoods.

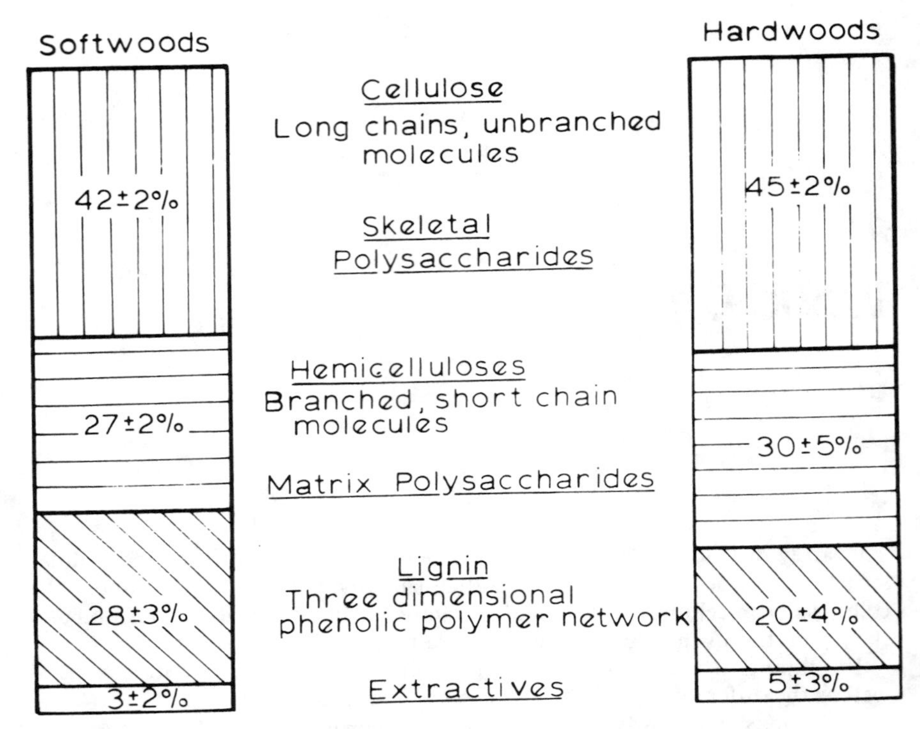

Figure 1.1 Average compositions of softwoods and hardwoods.

Cellulose

Cellulose is the chief part of the cell walls of wood.

Cellulose is a complex polymeric carbohydrate having the same percentage composition as starch, which yields glucose on complete hydrolysis by acid. It consists of a long chain of identical molecules bound together. These beta-glucosidic residues are linked through the 1, 4-positions.

The basic molecular unit in cellulose is glucose, which is a sugar. (See Figure 1.2.) Many glucose molecules are linked together to form a cellulose chain. (See Figure 1.3.)

The chemical formula for cellulose is $(C_6H_{10}O_5)_n$. Where n is the number of repeating glucose units, n is also called the degree of polymerization (DP). The value of n varies with different sources of cellulose. During pulping in a digester, the degree of polymerization will decrease to a certain degree. It is important that it does not decrease too much since shorter cellulose chains will ultimately result in weaker pulp.

Cellulose in wood has an average degree of polymerization of around 3500, while cellulose in pulp has an average degree of polymerization in the 600 to 1500 range. Cellulose is an unbranched, linear polymer. This makes it possible for several cellulose chains to pack together and form a very ordered crystalline structure.

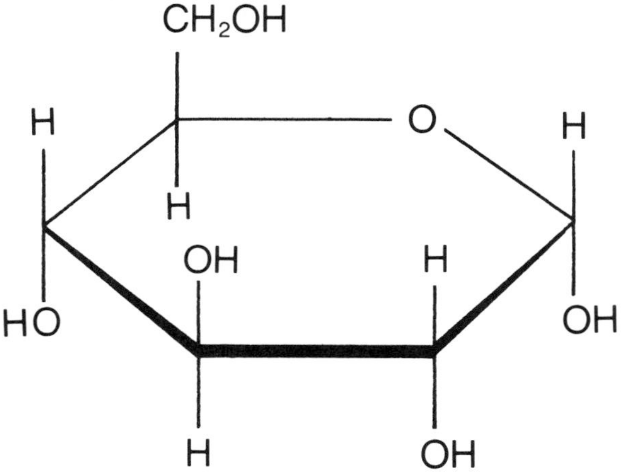

Figure 1.2 Glucose molecule.

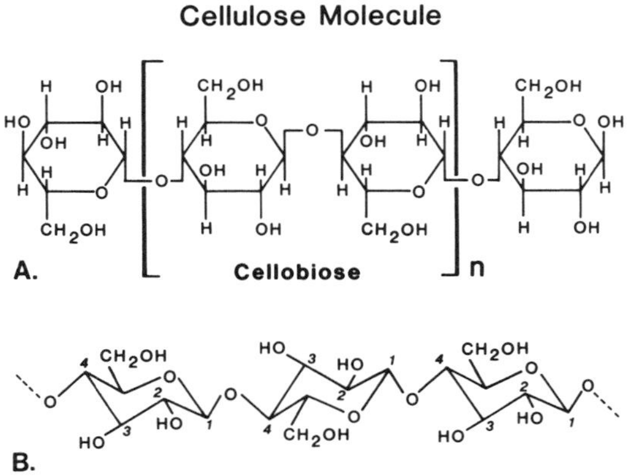

Figure 1.3 Cellulose chain.

Between the ordered crystalline regions, also called micelles, there are amorphous regions where the cellulose chains are not very ordered. (See Figure 1.4.) These bundles of aggregated cellulose chains are called microfibrils. Several microfibrils form a bigger structure, which is called a macrofibril. The fibrils pack together with hemicellulose, having lignin in between, to form the wood fiber wall. (See Figure 1.5.)

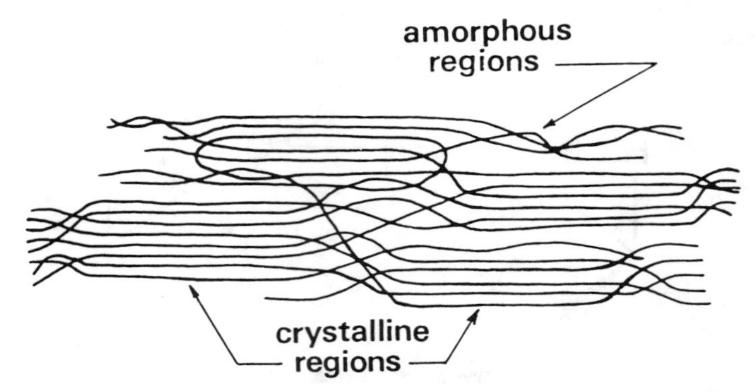

Figure 1.4 Cellulose chains forming crystalline and amorphous regions.

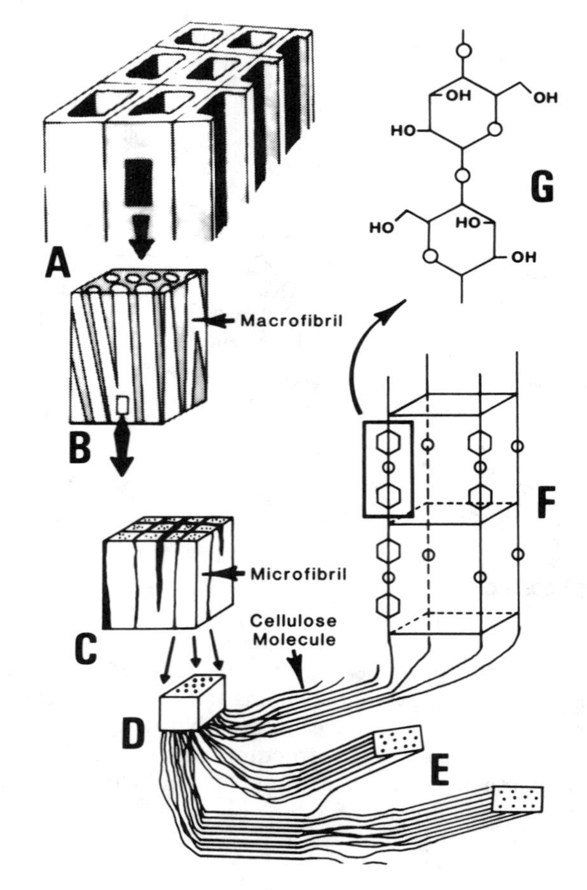

Figure 1.5 Cellulose structure in the fiber wall.

Hemicellulose

Hemicelluloses are also polymers mainly built of sugar units. In contrast to cellulose, which is a polymer of only glucose, the hemicelluloses are polymers of five different sugars: glucose, mannose, galactose, xylose, and arabinose. (See Figure 1.6.)

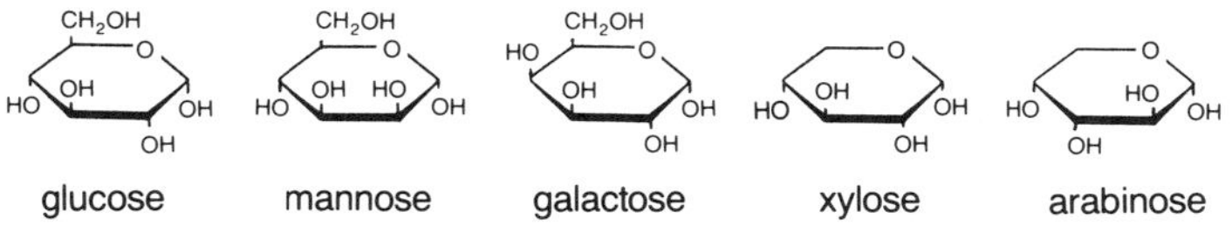

glucose mannose galactose xylose arabinose

Figure 1.6 Hemicellulose sugars.

There are different kinds of hemicellulose. (Figure 1.7) Different wood species contain hemicelluloses of slightly different composition. Hardwoods have more Xylan, while softwoods have more glucomannon. The type of hemicellulose also varies depending on the location within the wood structure.

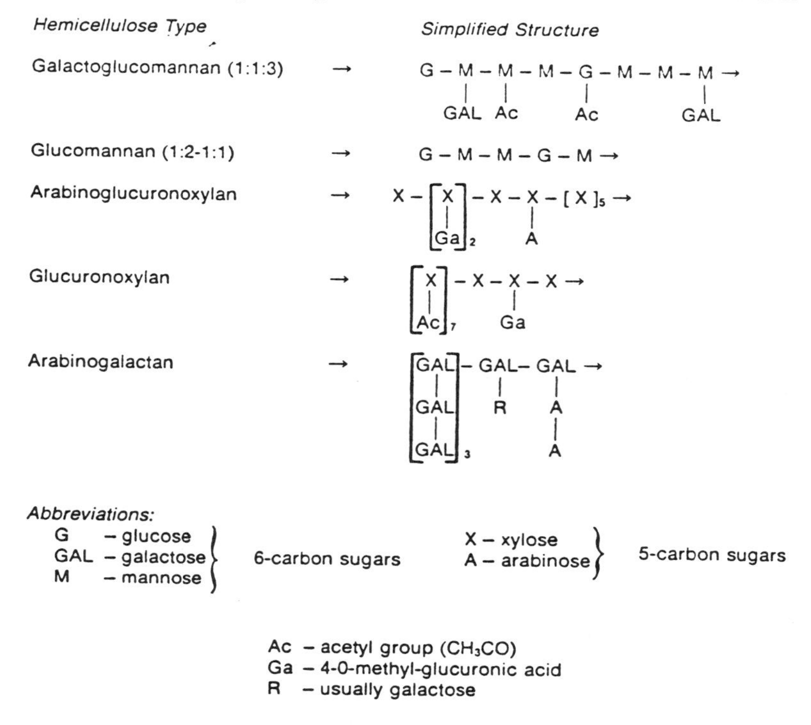

Figure 1.7 Different types of hemicellulose.

Hemicellulose chains are much shorter than cellulose chains; that is, they have a lower degree of polymerization. A hemicellulose molecule contains up to 300 sugar units. Also, in contrast to cellulose, hemicellulose is a branched polymer. This means it is not linear and cannot form crystalline structures and microfibrils as cellulose does. In pulping, hemicellulose reacts faster than cellulose.

In wood, the hemicelluloses are mostly found around the cellulose microfibrils, which they support. In papermaking, the hemicellulose will aid in making the paper stronger.

Lignin

Lignin is an amorphous substance that together with cellulose forms the woody cell wall of trees. It the cementing material between the cells and, thus, provides added mechanical strength to the wood.

Lignin is a highly branched, three-dimensional polymer. The basic molecular unit in lignin is phenyl propane. (See Figure 1.8.)

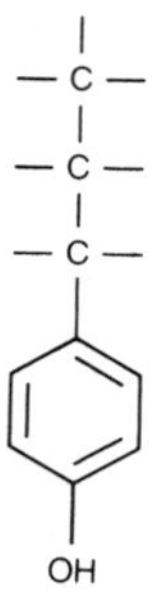

C
C
C
OH

Figure 1.8 Phenyl propane unit.

A lignin molecule is large with a high degree of polymerization. Because of its size and three-dimensional structure, the lignin in wood functions as a glue, or cement. (See Figure 1.9.)

The middle lamella, which mainly consists of lignin, holds the cells together and gives wood its structure. The cell walls also contain lignin. In the cell walls, lignin, together with hemicellulose, forms the matrix ("cement") in which the cellulose microfibrils are arranged.

Extractives

Wood usually contains a small amount of various substances that are called "extractives." These substances can be extracted from wood by either water or an organic solvent such as alcohol or ether.

Fatty acids, resin acids, waxes, terpenes, and phenolic compounds are some of the groups that make up the extractives. Most of these extractives are recovered in kraft pulping.

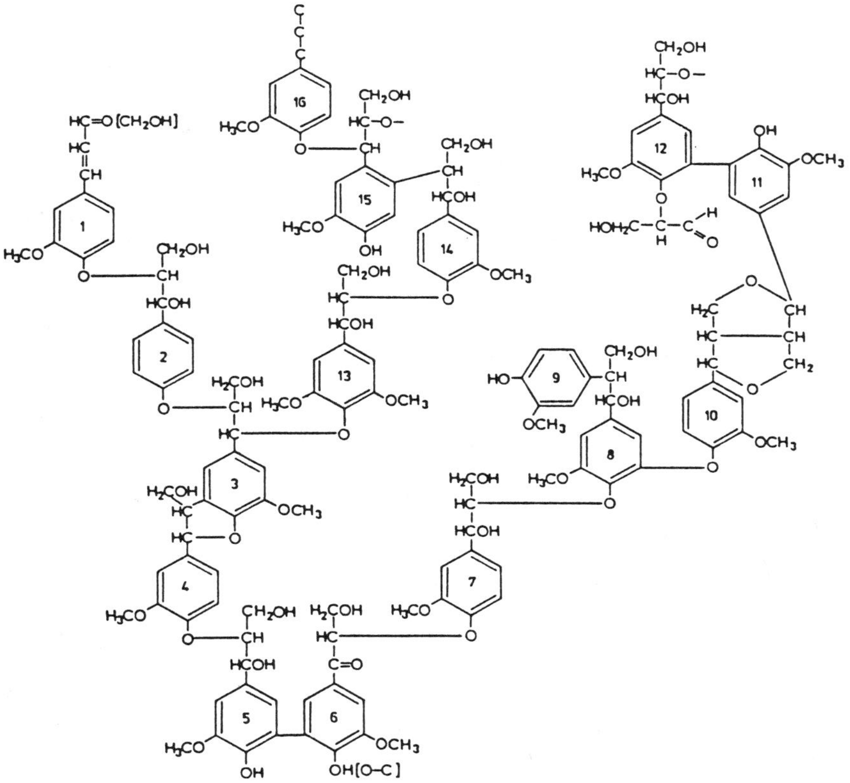

Figure 1.9 Native lignin.

Crude turpentine can be recovered from digester relief-gases. It has its origin in volatile terpenes. Fats, fatty acids, and resin acids are converted to soaps by the kraft process and dissolve in the cooking liquor. These soaps are later separated from the black liquor and recovered as tall oil. The tall oil is further purified and the products are used to manufacture different chemical products.

Some of the less soluble extractives can cause pitch problems in kraft pulping and in papermaking. Pitch consists of small aggregates of undissolved extractives. These aggregates can form sticky deposits on equipment, such as screens and paper forming wires. Pitch also appears as tiny specks in the manufactured paper.

Structure of a Tree Stem

Figure 1.10 shows a cross-section of a tree stem.

The *cambium* is a thin layer of cells between the bark and the wood where the tree growth takes place. It produces wood cells on the inside and bark on the outside. The rate of growth depends on the seasons. Springwood (or earlywood) is produced during the first part of the growing season, while summerwood (or latewood) is produced during the latter part. Because springwood and summerwood have a slightly different appearance, one can see the annual growth rings and, by counting them, determine the age of the tree.

The bark consists of two layers—the inner bark (also called the phloem) and the outer bark. The inner bark is a narrow layer of living cells through which the transport of sap takes place to provide the tree with energy. The outer bark consists of dead cells, which were once part of the inner bark. The outer bark contains a high fraction of extractives and its function is to protect the tree.

The outer, lightly colored section of the wood is *sapwood*. It gives structural support to the tree crown, acts as a food storage reservoir, and transports water from the roots to the tree crown. Some of the cells in the sapwood also transport sap.

The inner, dark-colored section of the wood is *heartwood*. It consists of dead cells that no longer have any function in transporting water or nutrients. Its only function is to provide support.

In most species the heartwood is much darker than the sapwood. This is because heartwood contains a higher amount of resins that are deposited in the heartwood's cell walls and cavities as the cells die.

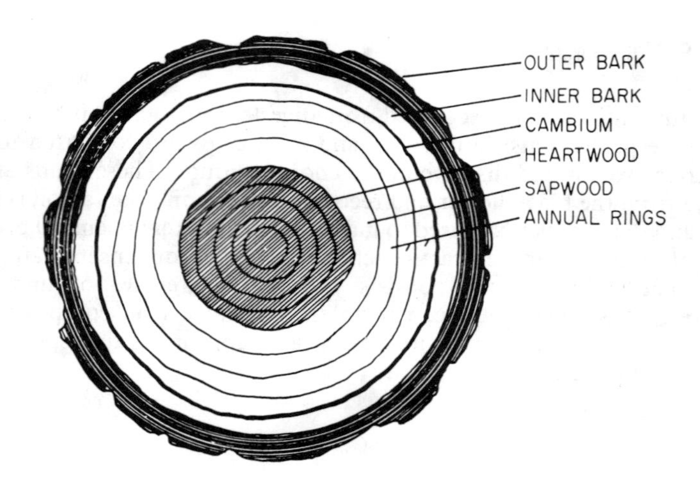

Figure 1.10 Cross section of a tree stem.

Cell Types in Wood

Wood is basically composed of long cells oriented along the stem direction. The cells are connected to each other through openings called pits. (See Figure 1.11.) Softwoods and hardwoods are built of partly different kinds of cells. Each species contains different cell types with each having special functions.

Cell types in softwoods

Figure 1.12 shows the cellular structure of a small section of a softwood.

The wood substance in softwoods is composed of two different cell types: *tracheids* (90% to 95%) and *ray cells* (5% to 10%). Tracheids are long, narrow cells oriented vertically along the tree stem. They give the tree its mechanical strength and provide for water transport, mainly through the thin-walled springwood-tracheids with their large cavities. The tracheids are 2 mm to 4 mm long but only 0.02 mm to 0.04 mm wide.

There are two different kinds of ray cells—*ray parenchyma* and *ray tracheids*. Groups of these cells, called rays, are oriented in a radial fashion, lying horizontally in the stem. The ray parenchyma are living cells that store and transport nutrients. The ray tracheids transport water in the radial direction. Ray cells are much smaller than vertical tracheids.

Most softwoods also contain *resin canals*. They consist of horizontal and vertical intercellular spaces building a uniform channel network in the tree. The resin canals are lined with a special kind of parenchyma cells called *epithelial* cells, which secrete resin.

Cell types in hardwoods

Hardwoods contain a greater number of cell types than softwoods. The most important cell types are *libriform cells, fiber tracheids, vessel elements* and *parenchyma cells.*

The libriform cells function as supporting tissue. They are relatively long cells with a length of 1 mm to 1.5 mm and a width of 0.01 mm to 0.04 mm.

The vessel elements are thin-walled and rather short, having very wide cells open at each end. Vessel elements are placed on top of one another to form long "tubes" through which water and nutrients are transported. Vessel elements range in size from 0.2 mm to 1.3 mm in length and 0.02 mm to 0.3 mm in diameter.

The parenchyma cells can be either vertical or horizontal. Their function is to store nutrients, and they are a high source of extractives and pitch. Hardwoods contain more parenchyma cells than softwoods. The rays in hardwoods contain no tracheids, only ray parenchyma cells.

The fiber tracheid is a kind of cell that mostly resembles the libriform cells. The term "hardwood fibers" includes both libriform cells and fiber tracheids.

Papermaking fibers in hardwoods and softwoods

Figure 1.13 shows all the major cell types in softwoods and hardwoods.

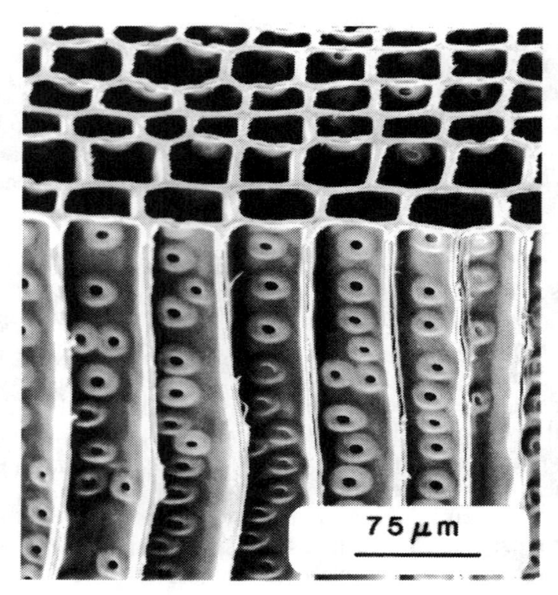

Figure 1.11 Wood fibers with pits.

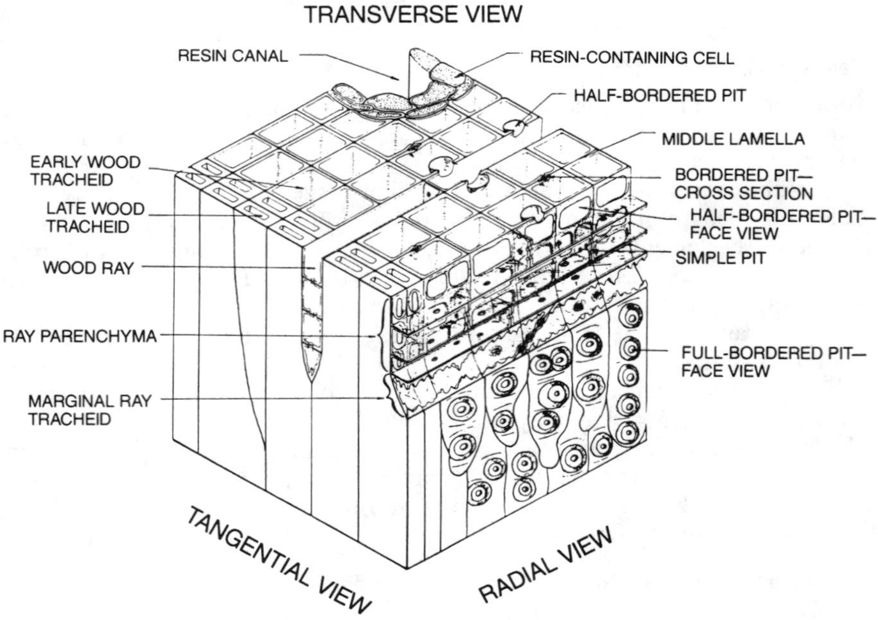

TRANSVERSE VIEW
RESIN CANAL
RESIN-CONTAINING CELL
HALF-BORDERED PIT
MIDDLE LAMELLA
EARLY WOOD TRACHEID
BORDERED PIT—CROSS SECTION
LATE WOOD TRACHEID
HALF-BORDERED PIT—FACE VIEW
SIMPLE PIT
WOOD RAY
RAY PARENCHYMA
FULL-BORDERED PIT—FACE VIEW
MARGINAL RAY TRACHEID
TANGENTIAL VIEW
RADIAL VIEW

Figure 1.12 Softwood cellular structure.

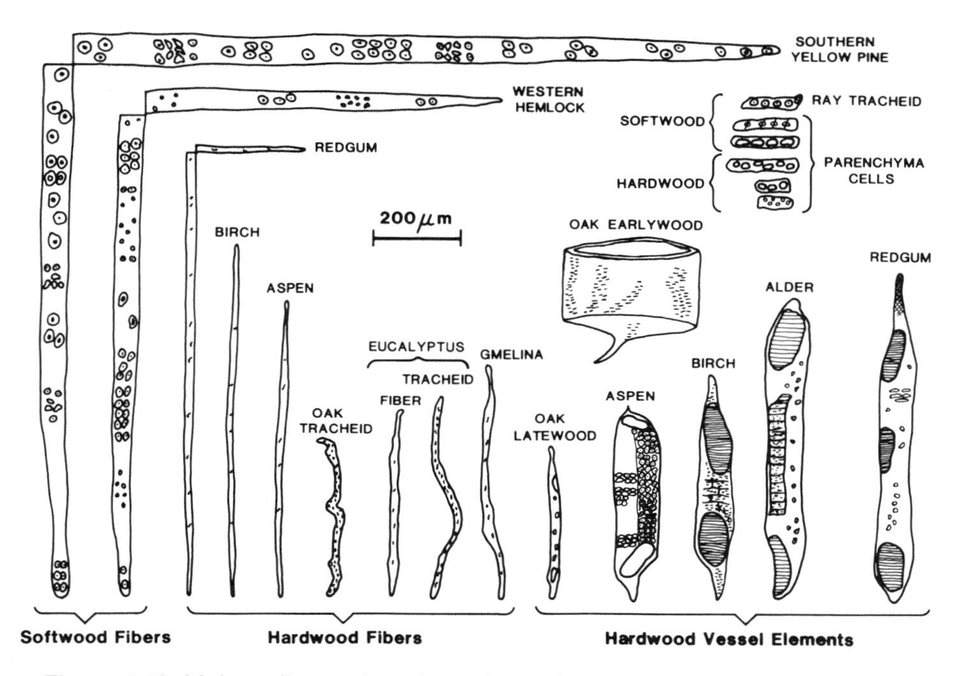

Figure 1.13 Major cell types in softwoods and hardwoods.

The papermaking fibers in softwoods are the vertical tracheids. In hardwoods the papermaking fibers consist mainly of libriform cells plus the fiber tracheids.

The major difference between softwood and hardwood fibers is the fiber length. Hardwood fibers are about one-third to one-half of the length of softwood fibers. Another difference between softwoods and hardwoods is the amount of the different cell types. Softwoods contain a higher fraction of "fibers" than do hardwoods. Figure 1.14 compares a softwood species (spruce) with a hardwood species (birch).

Types of cells — spruce vs. birch

	fibers (%)		vessels (%)		parenchyma (%)	
	by wt.	by vol.	by wt.	by vol.	by wt.	by vol.
spruce	99	95	—	—	1	5
birch	86	65	9	25	5	10

Figure 1.14 Amount of different cell types in spruce and birch.

Parenchyma cells in both softwoods and hardwoods are so small that they are usually almost totally degraded in the pulping and bleaching operations. If not, they can contribute to pulp fines. The parenchyma cells are mainly responsible for the generally higher fines content of hardwood pulps. The parenchyma cells are also the main source of pitch.

Hardwood vessel elements vary in shape. The longer, narrower vessel elements in some species may end up as papermaking fibers, while the large diameter vessel elements of some species, i.e., oaks, are troublesome in papermaking.

Structure of a Softwood Tracheid

Figure 1.15 shows the cross-section of a softwood tracheid. The cell wall is composed of several layers. First comes the relatively thin *primary wall* (P), then the thick *secondary wall* divided into three sections: S_1, S_2, and S_3. Sometimes the S_3 section is called the *tertiary wall*. The inside of the S_3 layer is covered with a thin membrane called the warty layer. The S_2 layer forms the main body of the fiber and is 2 μm to 10 μm thick. The fiber is hollow, and this central cavity is called the *lumen*.

The cellulose microfibrils are oriented in different directions in the different fiber wall layers. In the primary wall they form an irregular pattern while the different layers of the secondary wall contain microfibrils oriented in helical patterns. The individual fibers are bonded together by the *middle lamella*, which consists mostly of lignin and some hemicellulose.

Lignin and hemicellulose can also be found in the fiber walls. As a matter of fact, a major part of the total lignin is located in the fiber wall, mainly in the S_2 layer. This is because the fiber walls take up such a big fraction of the wood compared to the middle lamella. Chemical pulping removes lignin from between the fibers, and from within the cell wall.

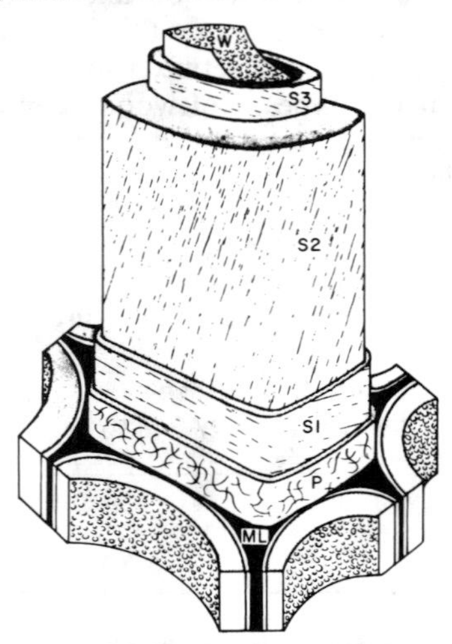

Figure 1.15 Cross-section of a softwood tracheid.

Questions

1. Name the major chemical components of wood.
2. Specify the approximate chemical composition of wood.
3. Understand the meanings of the terms "polymer" and "degree of polymerization."
4. Discuss the chemical structure of cellulose.
5. Describe the structure of a cellulose microfibril.
6. State the main differences between cellulose and hemicellulose.
7. Discuss the chemical structure of lignin and its role in the pulping process.
8. Describe where lignin is present in the wood structure.
9. List a few of the organic compounds making up the extractives fraction of wood.
10. Name two kraft pulping by-products originating from the extractives fraction of the wood.
11. Explain how extractives can cause problems in pulping and papermaking.
12. Name and identify the different types of wood and bark appearing in a cross section of a wood stem.
13. Describe how the growth of a tree takes place.
14. Name the two main types of wood cells in softwood trees.
15. Give the approximate size of a softwood tracheid.
16. Identify what specific function the different cell types have in a softwood tree.
17. List four types of hardwood wood cells.
18. Identify what specific function the different cell types have in a hardwood tree.
19. Specify what wood cells are considered as papermaking fibers in softwoods and hardwoods.
20. Discuss the impact of parenchyma cells on pulping and papermaking.
21. Describe the general structure of a softwood tracheid.
22. Name the different layers in the cell wall of a softwood tracheid.

Principal Sources for Figures

Kocurek, M. J., and Stevens, C. F. B., eds. *Pulp and Paper Manufacture.* Vol. 1: *Properties of Fibrous Raw Materials and Their Preparation for Pulping.* 3rd ed. Atlanta and Montreal: Joint Textbook Committee of the Paper Industry, 1983.

Smook, G. A. *Handbook for Pulp and Paper Technologists,* Edited by M. J. Kocurek. Atlanta and Montreal: Joint Textbook Committee of the Paper Industry, 1982.

Fig. 1.1 Smook, Fig. 2.18, p. 15.
Fig. 1.3 Kocurek, Fig. 43, p. 35.
Fig. 1.4 Smook, Fig. 1.5, p. 5.
Fig. 1.5 Kocurek, Fig. 44, p. 36.
Fig. 1.7 Kocurek, Fig. 49, p. 39.
Fig. 1.9 Kocurek, Fig. 50, p. 41.
Fig. 1.10 Smook, Fig. 2.2, p. 10.
Fig. 1.11 Kocurek, Fig. 24, p. 24.
Fig. 1.12 Smook, Fig. 2.3, p. 10.
Fig. 1.13 Kocurek, Fig. 40, p. 32.
Fig. 1.14 Smook, Fig. 2.2, p. 15.
Fig. 1.15 Smook, Fig. 2.5, p. 12.

2
Chip Quality

The quality of the chips used for pulping is a very important factor in the operation of the pulp mill and in final pulp quality. Therefore, it is important to know what variables will affect the chip quality and the effect they have on the pulping operation and pulp quality.

Chip quality variables can be divided into two sections: wood-related variables and process-related variables. Wood-related variables are concerned with properties of the wood itself, such as selection of species, variation within species, wood deterioration during chip storage, and wood decay.

Chip quality variables related to the chipping operation, chip size distribution, and chip bulk density include blade sharpness, chipper design, screen type and operation.

Wood-related Variables

Wood species

Hardwood and softwood

Generally, softwood chips produce a stronger pulp than hardwood chips. This is because the softwood fibers are longer and more flexible than hardwood fibers.

Softwoods normally give a slightly lower yield than hardwoods when pulped under the same conditions. This occurs because softwood hemicellulose is more soluble than hardwood hemicellulose, and softwood generally contains more lignin than hardwood.

Hardwood pulps produce a paper with good printing qualities. The hardwood fibers form a smooth surface because of their small size.

Wood density

Wood density is a very important economic factor in pulping. With a denser wood, one can pack more weight into a given digester volume and thus increase pulp production, either per batch cook or per time unit in a continuous digester.

A more commonly used parameter than density is specific gravity. Specific gravity is defined as the ratio of the weight of a wood sample to the weight of water displaced by that same sample. Since both weight and volume of wood changes with moisture content, it is more convenient to use a modified parameter based on dry-wood weight and water-swollen volume. This parameter is called basic specific gravity (BSG) and is defined as:

$$\text{BSG} = \frac{\textit{oven-dry wood weight}}{\text{weight of water displaced by the same wood sample in its water-swollen state.}}$$

BSG is a ratio and thus unitless.

Basic density is a related parameter defined as:

$$\frac{\textit{oven-dry weight}}{\text{green volume}} \qquad \text{units : } \frac{g}{cm^3} \text{ or } \frac{lbs}{ft^3}$$

Wood specific gravity varies with wood species. Hardwoods tend to have higher specific gravities than softwoods. Some examples of specific gravity for different species can be seen in this table:

Species	Specific Gravity
Douglas fir	0.41 to 0.45
White spruce	0.34 to 0.37
Western red cedar	0.29 to 0.31
Red pine	0.39 to 0.44
Red maple	0.44 to 0.50
White oak	0.59 to 0.67
Birch	0.48 to 0.55

Specific gravity varies within species and within the same tree as well. This variation depends on cell wall thickness and cell size. Wood with thicker cell walls has more wood substance and less void space per unit volume than wood with thin cell walls. Therefore, it has a higher specific gravity.

Pulp quality and later operations, such as washing refining, and papermaking, are affected by the specific gravity of the wood source. Thick walled fibers (high density wood) produce a coarse pulp with stiff fibers and a high water drainage rate. Thin-walled fibers (low-density wood) produce a pulp with very flexible fibers that give a high strength and high-paper density.

Species used for bleached pulps

Bleached kraft hardwood pulps are commonly used for fine paper and high quality publication grades where the main requirements are good sheet formation and a good printing surface. High strength is not the primary concern. The most commonly used species are aspen, birch, maple, oak, red gum, and tupelo.

Bleached kraft softwood pulps are used extensively for packaging papers where high strength is critical. The most commonly used species are spruce, pine, hemlock, larch, and Douglas fir.

Softwood pulps are often mixed with hardwood pulps to give the stock higher strength, primarily to withstand the tensions existing in high speed printing presses.

Species used for linerboard

Linerboard is used as the outer plies of corrugated box stock and as wrapping paper. It consists of two layers called the top and the base liner. The top liner is cooked to a lower yield to provide a smoother surface for printing.

Linerboard must be tough and strong with a high level of stiffness and burst resistance. The top liner also must have good printability. Because of the strength requirements, linerboard is primarily produced from softwood pulp. However, a quantity of hardwood pulp can be used for the top liner to improve surface smoothness and printability.

Variation within species

Growth location

Wood quality is affected by the environment where the tree grew. Climate, soil fertility, and elevation are the main factors.

For softwood, the properties altered most noticeably by the environment are ring width, earlywood/latewood ratios, fiber wall thickness, and fiber size. As a general rule, warm temperature and adequate moisture combine to produce wood of a higher specific gravity.

Juvenile and mature wood

A tree does not produce exactly the same type of cells over its whole life. Therefore, there is a difference between juvenile wood produced during the first 5 to 20 years of a tree's lifetime and the mature wood produced thereafter. The differences are more marked in softwoods but are still noticeable in hardwoods. Generally, compared to mature wood, juvenile wood has these characteristics:

- Shorter and narrower fibers and vessel elements
- Thinner cell walls
- High earlywood/latewood ratio
- Lower specific gravity
- Lower cellulose content
- Higher hemicellulose content
- Higher lignin content
- Wider growth rings
- Higher fraction reaction wood.

Because of the different properties of juvenile and mature wood, chips from young trees will not have the same pulping properties as chips from a fully

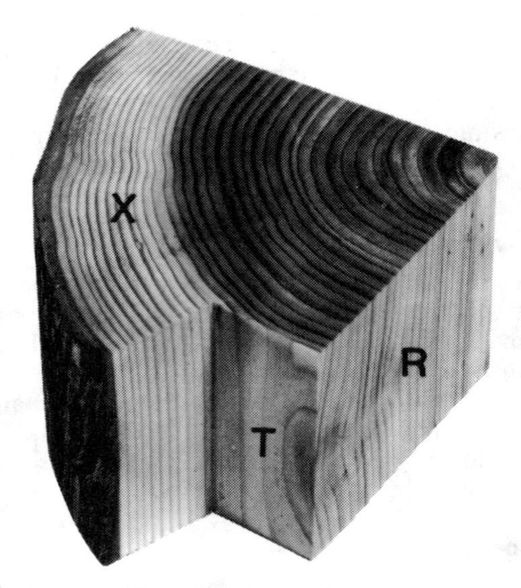

Figure 2.1 Photo of stem with early/latewood.

grown tree. A big disadvantage of juvenile wood kraft pulping is a lower yield and an increase in alkali consumption. The reason for this is the lower cellulose content.

Earlywood and latewood

The growth rings in a tree stem consist of two sections: a lighter, wider section that is formed during the early growing season, and a darker, more narrow section that is formed during the late growing season. (See Figure 2.1.)

The wood produced during the early part is called earlywood (or springwood). It consists of thin-walled fibers of large diameter. The wood produced during the later part of the growing season is called latewood (or summerwood). It consists of more thick-walled fibers with a narrower radial diameter. Figure 2.2 shows earlywood and latewood fibers in cross-sections of wood.

The ratio of earlywood to latewood is an important factor in pulping. Earlywood has a lower specific gravity than latewood and the thin-walled earlywood fibers collapse more easily to form flat ribbons that provide a large area for fiber bonding during papermaking. Thick-walled latewood fibers retain their tubular form and, therefore, do not bond as well. A decreased earlywood to latewood ratio means a higher specific gravity and decreased interfiber bonding (a weaker paper sheet) unless process changes are made, for example, increased refining before papermaking.

Wood grown in more northern latitudes or at higher altitudes, where the growing season is shorter, contains a higher ratio of thin wall fibers. Therefore, sheets made of northern kraft pulps tend to have higher tensile and bursting

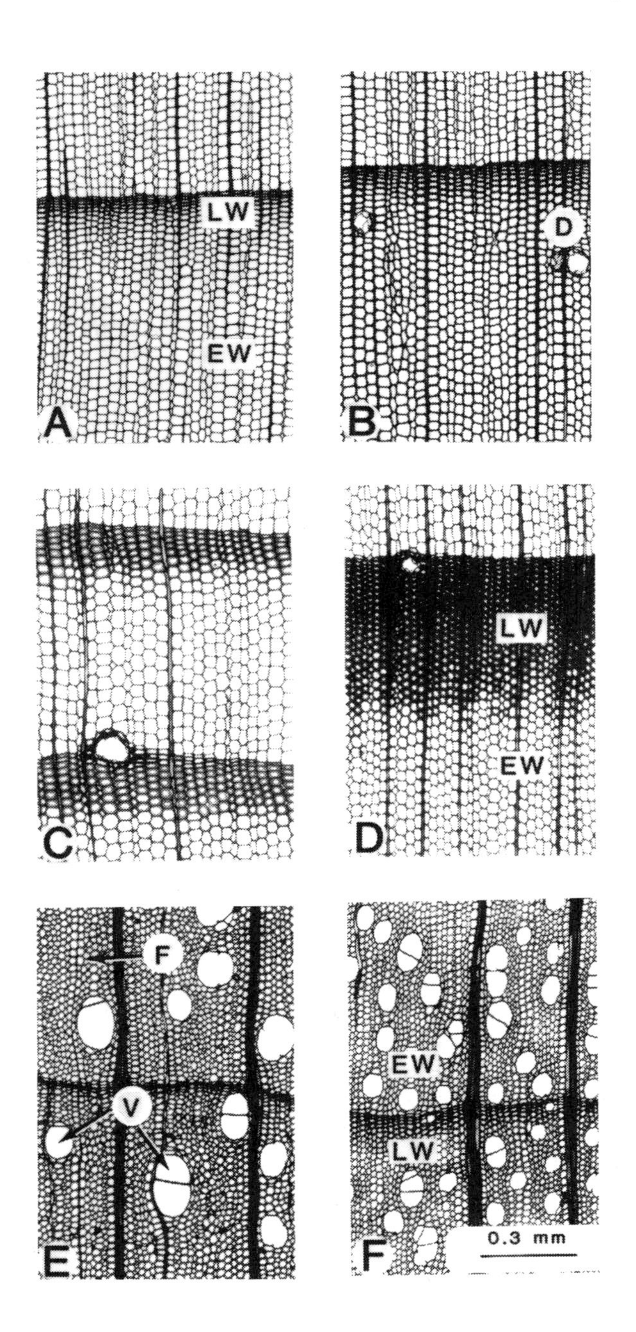

Figure 2.2 Cross-sections of wood showing earlywood and latewood.

strengths but lower bulk, porosity, and tearing resistance than sheets made of southern kraft pulps.

Reactionwood

Reactionwood is wood formed in a stem that is leaning or where it is branched. Its function is to help the tree maintain a vertical stem and keep the branches oriented in a preferred direction.

Reactionwood in conifers is called *compression wood.* It is located on the underside of a leaning stem and under a tree branch. Reactionwood in hardwoods is called *tension wood.* It is located on the upper side of leaning stems and tree branches.

Compression wood has a darker color than normal wood. Its fibers are also shorter and more thick-walled. Compression wood also has a higher specific gravity, higher lignin content, and lower cellulose content than normal wood. Because of these factors, compression wood gives lower yields than normal wood. Also, the pulp is darker and has lower strength properties.

Tension wood has a lighter color than normal wood. Fiber length is not affected, but the fiber walls are thicker than normal. The specific gravity is higher, cellulose content higher, and lignin content lower than for normal wood. Yields and brightness from tension wood are somewhat higher because of the higher fraction cellulose. Strength properties are lower than for normal wood since the thick-walled cells result in lower fiber bonding.

Sapwood and heartwood

The wood in young trees and the outer wood in older trees contains living cells and can transport sap. It is, therefore, called *sapwood.* The wood in the middle of an old tree consists of dead cells and is not involved in transporting sap. It is called *heartwood.* Normally, sapwood is light in color while heartwood has a darker color, especially in hardwoods. When heartwood develops, extractives fill cell voids and impregnate the fiber walls. Therefore, heartwood is not very easily penetrated by liquids.

In pulping, heartwood can cause problems. It does not chip as easily as sapwood because of its lower moisture content and the higher fraction of extractives. Also, the cooking liquor does not penetrate as easily. This can cause large amounts of knotter and screen rejects. The color of heartwood can also cause a problem since the colored extractives must be removed to increase the pulp brightness.

Wood source

Roundwood

Prior to 1950, almost 100% of all chips were prepared from roundwood; that is, from the entire stem of a tree. Since then, there has been a considerable change in sources of wood for pulping.

- Today about 50% of the chips originate from roundwood.
- Chips produced from roundwood are normally of good quality.
- Roundwood chips are typically chipped at the pulp mill.

Sawmill residuals

In the 1950s, with increasing timber prices, pulp mills started to look for other chip sources than roundwood.

Earlier, sawmills had burned or dumped the slabs that were produced when sawing lumber. Then they started to debark the logs before converting them into lumber. The slabs and waste wood produced were then run through a chipper, and the chips were sold to a nearby pulp mill.

Chips from sawmill residuals contain a high fraction of sapwood, since the origin is the outer parts of the tree stem. Slabs from older trees consist of mature wood.

An efficient debarking operation is very important when chipping residual chips. After sawing a debarked log, about 40% of the wood remains as slabs, which can be chipped into chips. Since it is the outer parts of the log that form the slabs, all bark remaining after the debarking will be converted into chips as well. An ineffective debarking operation will, therefore, leave a much higher percentage bark in residual chips than it would in roundwood chips. Most residual chips are chipped at the sawmill. It is important that the pulp mill know of any changes in debarking and chipping conditions since it will affect the chip quality.

Whole-tree chipping

Whole-tree chipping is a term that covers a variety of practices, from chipping and pulping every part of a tree including roots, stem, branches, and leaves, to chipping only selected parts such as stems and branches. The common denominator for these practices is that the tree is converted to chips in the forest on or near the logging site.

The quality of whole-tree chips varies considerably depending on what kind of wood and what parts of the trees are chipped. Large oversize and undersize fractions, high content of bark, rot, and contaminants, such as sand and dirt, are some of the negative quality variables that must be considered. Whole-tree chips, therefore, need to go through more extensive cleaning than other chips.

Whole-tree chips are usually considered to have a lower quality profile than residual or roundwood chips. Therefore, they are mostly used in coarser grades of paper such as linerboard.

When pulping whole-tree chips one must expect a higher alkali consumption due to a higher bark content. Pulping whole-tree chips will also result in a slightly lower yield and pulp strength than when pulping roundwood chips.

Chip storage

Reason for chip storage

The basic reason for chip storage is that a pulp mill must have a large enough chip inventory to be able to supply chips to the process at all times.

If the mill chips its own chips from roundwood, the size of the chip inventory can be easily controlled and regulated. If the mill depends on delivered (residual) chips, though, a higher chip inventory will be necessary. Basically, this is because the pulp mill does not have total control over chip delivery. Also, economic factors, such as low residual chips prices combined with a bad market for pulp, can cause the chip inventory to grow considerably.

In general, mills maintain two to six weeks' chip storage, but inventories may be as large as one year's supply for a mill. Normally, the chips are stored in outdoor piles. In addition, most pulp mills have chip bins or silos that will hold from several hours to a few days' chip supply.

Loss of wood and by-products

Chips stored in piles deteriorate through several degrading reactions. Microorganisms attack the chips, the extractives are broken down by oxidation, and cellulose and hemicellulose are hydrolyzed to smaller molecules. A wood loss of about 1% per month is typical. The extractives are lost at a much faster rate than that, though. At a pile temperature near 50°C, tall-oil losses of 50% in the first month can be expected.

Turpentine is lost even more rapidly at 50°C, and may be completely absent after three months.

Other negative effects

A lot of heat is generated in a chip pile because of all the chemical and biological reactions that are taking place.

The increased temperature speeds up the ongoing chemical reactions even further. If the temperature does not stabilize, but rises above 80°C, the pile can catch fire.

The content of fines usually increases following outside chip storage. During storage the chips become more brittle because of the biological and chemical reactions and, when handled, fines will break off. Fines will reduce the overall pulp yield.

When the cellulose and hemicellulose have been degraded by acid hydrolysis the loss in pulp yield is considerable even though the chips may not have lost much weight. This occurs because acid hydrolysis splits the carbohydrate chains to shorter chains and these will dissolve during digestion.

A loss in pulp quality also occurs when pulping chips have been stored for a longer time. Both strength and brightness of the pulp are decreased.

Chip pile management

Because of the harmful changes that can happen in chips during storage, it is important to manage the chip storage properly. Many mills try to reduce the size of their inventory to minimize storage time.

Some guidelines for outside storage of chips are:

- Follow the practice of first-in-first-out. This way, all chips will be in storage for about an equal length of time.
- Minimize the height of the chip pile, because the temperature is highest at the top of the pile. Keep pile height below 15 m.
- Monitor pile temperatures routinely.

- Avoid production of fines and compacted chips. Restrict driving tractors on top of the pile as much as possible.
- Use tractors with rubber wheels.

Wood decay

Origin of decay

Wood decay is caused by different kinds of microorganisms—mainly fungi, but also bacteria, yeasts, and molds.

Decay occurs both in standing, living trees and in piles of stored chips or pulpwood logs. These two decay situations should be looked upon separately because they are caused by different organisms. Wood decay will always occur to some degree during chip storage. Distinct from this is the wood decay that takes place in some living trees.

White and brown rots

White and brown rots are both groups of fungi that attack standing trees.

Brown rots are more common in softwoods but do attack hardwoods as well. (See Figure 2.3.) In advanced stages of decay, the wood turns brown and can be easily broken by hand. All brown rots prefer carbohydrates over lignin and extractives. The fungi attacks the cellulose and hemicellulose by random cleavage of the chains. This means the degree of polymerization (that is, the chain length) decreases quickly. Therefore, the loss in pulp strength is big even at early stages of decay.

White rots generally attack both lignin and carbohydrates, leaving a whitish wood color behind. The white rot fungi attack the cellulose by "peeling" the glucose units at the end of the chain. The chain length is, therefore, reduced rather gradually and, as a result, white rots do not reduce pulp strength as rapidly as brown rots.

Effects of decay on pulping

Decayed wood can be processed and pulped in the kraft process, but there are many factors that reduce the profitability.

First, chipping and chip handling of decayed wood is more difficult. More fines and over/undersize chips are produced. Decayed wood has a lower density, which means less pulp can be produced per batch cook or per time unit in continuous digesting. Also, the weight percent yield decreases because of the degraded cellulose in the decayed wood. Yield reductions up to 25% have been reported. This causes the alkali consumption to increase as well. The decayed wood contains a higher percentage of lignin and carbohydrates that are made more susceptible to chemical breakdown by the pulping chemicals than sound wood and will, therefore, consume more chemicals. Increases in alkali consumption up to 25% per ton pulp have been recorded.

The degradation of cellulose also leads to a reduction in pulp strength, as mentioned earlier.

Figure 2.3 Brown rot in Douglas-fir.

The lower yield and increased chemical consumption mean a higher load on the recovery boiler. Since the capacity of the recovery boiler often is the limiting factor in kraft pulp mills, the mill productivity will be lower when pulping decayed wood than under normal circumstances.

Process-Related Variables

Chip size distribution

Chip thickness

In kraft pulping, chip thickness is the most important size parameter determining chip quality. This is because the kraft cooking liquor can penetrate the chip at about the same rate from all directions. Therefore, thickness is the dimension that determines when the chip is fully penetrated.

Laboratory work has shown that 3 mm is the optimal thickness for hand-cut laboratory chips. For technical chips, which contain more cracks and fissures that help liquor penetration, the optimal thickness is about 5 mm.

If a chip is too thick, the cooking liquor will not have time to fully penetrate the center of the chip during cooking. This uncooked center will end up as a shive or a knot in the rejects.

If one tries to compensate for the chip thickness by increasing the cooking time or liquor concentration, the result will be a fully cooked center with the outer parts of the chips overcooked, resulting in lower pulp yield and decreased pulp strength.

Oversize and undersize fractions

Oversize chips (see Figure 2.4) are generally overthick as well and will, therefore, not cook uniformly, giving a high level of rejects. Oversize chips can also cause non-uniform chip packing in the digester. This will lead to non-uniform liquor circulation and uneven cooking.

Figure 2.4 Photo of oversize chips.

Figure 2.5 Photo of pinchip that has split off from a regular size chip.

Pin chips are chips that have approximately the same width and thickness. (See Figure 2.5.) A large fraction of pin chips can cause liquor circulation problems and even plugging, especially in continuous digesters. The tolerance limit for pin chips is about 10% to 12% in regular chips.

The undersize fraction consists of fines and sawdust. This fraction also causes problems with maintaining uniform liquor circulation and risks for plugging. Continuous digesters are the most sensitive since the undersize material tends to plug their liquor circulating screens. Only a very small fraction, 1.0%, can normally be tolerated in chips.

The undersize fraction will normally be overcooked when digested with regular size chips. Therefore, pulp yield and pulp strength are lower for this fraction. However, the main reason for the lower pulp strength is that the fibers in the sawdust and fine particles are so short and physically damaged.

The undersize fraction can be digested separately in special digesters. With a cooking time somewhat shorter than for regular chips, a normal yield can be obtained. However, the strength properties are clearly inferior compared to pulps made from normal chips. Figure 2.6 illustrates the lower strength properties of sawdust pulps.

Evaluation of size distribution

A narrow size distribution and uniform chip thickness are very important to assure uniform liquor penetration and even cooking conditions. Therefore, a good method to evaluate size distribution on a lab scale basis is needed to run tests on chips purchased or chipped at the mill.

The chip size distribution is evaluated with some kind of screening system, usually called a classifier. To be able to evaluate chip thickness, one should use a classifier that contains a slotted screen that catches overthick chips.

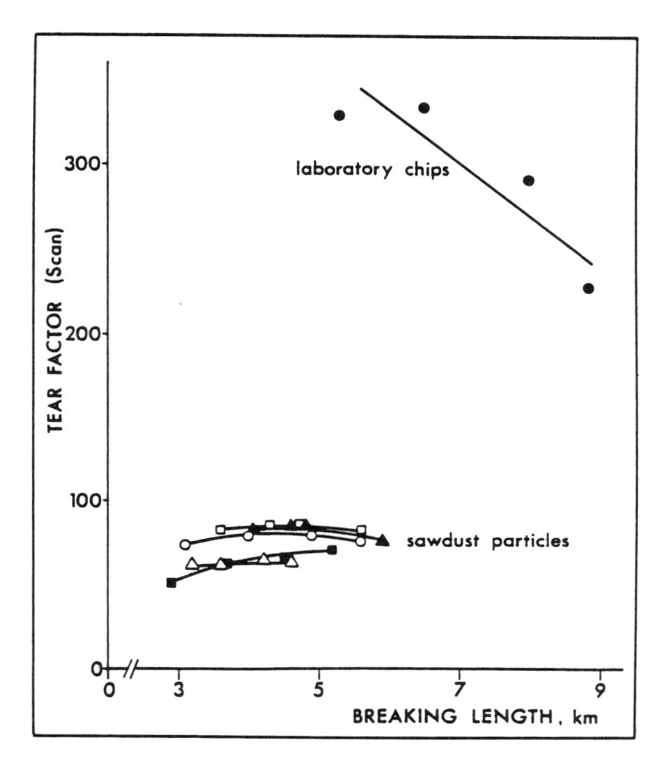

Figure 2.6 Tear factor and break length for sawdust and chips.

Figure 2.7 shows a chip screening system for laboratory use.

- The top screen has 45 mm wide round holes. The chips retained on this screen are called *overlarge.*
- The next screen has 10 mm wide slots for softwoods (8 mm for hardwoods). The chips retained on this screen are *overthick.*
- The third screen has 7 mm round holes. On this screen the *accepts* fraction is retained.
- The fourth screen with 3 mm round holes will retain the *pin chips.*
- What passes all four screens and collects on the bottom is the fines fraction, or *pan fines.*

The following is a typical result of classified roundwood chips:

+45 mm overlarge	2%
+10 mm overthick	4%
+ 7 mm accepts	85%
+ 3 mm pins	7%
pan fines	2%
	100%

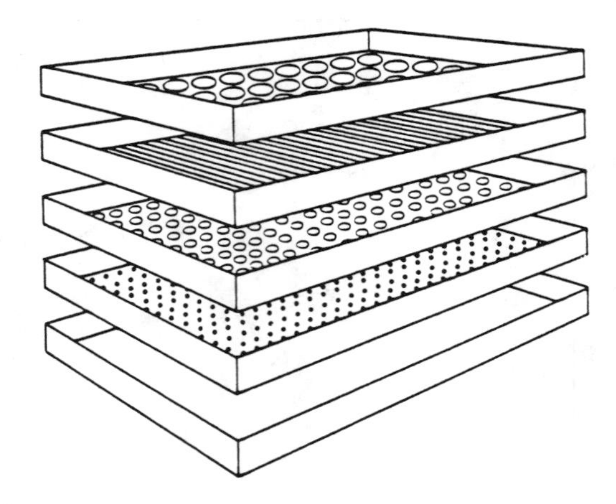

Figure 2.7 Chip screening system.

Chip bulk density

Effect of chip bulk density

Chip bulk density is an important parameter in chip handling or when filling a digester. It determines the amount of wood that can be charged. It also influences the filtration resistance when liquor is circulated through the chips in the digester.

Bulk density is usually expressed in lb/ft^3 (or kg/m^3) on either a dry-wood or a wet-wood basis. The chip bulk density is affected by wood density, chip size, and chip size distribution. The filling mechanism, if some kind of compacting is used, also affects the final bulk density in the digester.

The chip size parameter that has the most control over the bulk density is the ratio of the largest to the smallest dimension taken as the diagonal of the chips divided by the chip thickness. (See Figure 2.8 & Figure 2.9.)

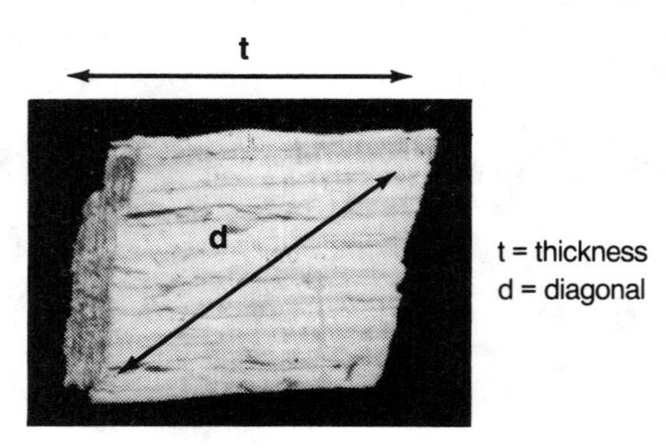

Figure 2.8 Thickness and diagonal of a chip.

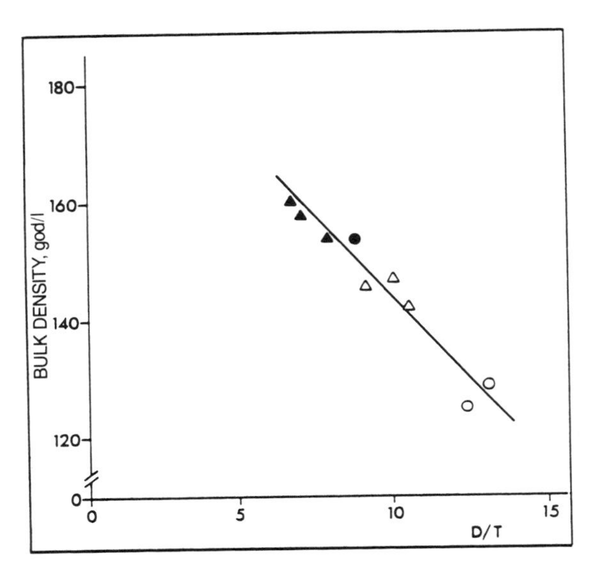

Figure 2.9 Bulk density vs. size ratio.

Measuring chip bulk density

Chip bulk density is determined through a standard laboratory procedure.

A cylinder of a given volume is filled with chips under controlled conditions. The chips are then weighed and the bulk density calculated. This procedure is illustrated in Figure 2.10.

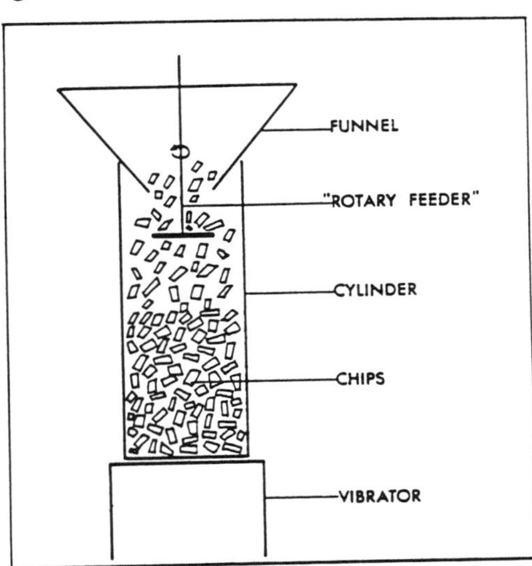

Figure 2.10 Laboratory apparatus for chip bulk density measurement.

Chip moisture

Effect of chip moisture

Chip moisture does have an effect on pulp yield, kappa number, and pulp quality. If the chip moisture is very low, it can be difficult to fully impregnate the chips.

It is important to know what the moisture content is to be able to calculate the true wood input to the digester and to keep alkali charge and liquor concentration constant from time to time. Moisture content of wood chips is normally expressed as a percentage of the wet weight (total weight) of the chips. Moisture levels are usually around 40% to 60%.

Measuring chip moisture

There are many ways of measuring chip moisture. If a mill uses purchased chips, the moisture content is normally determined at the time of delivery. Samples are taken from the load, weighed, and dried in an oven to establish the moisture content.

There is also equipment that can measure moisture content on-line, automatically and continuously. There are moisture sensors based on different principles such as capacitance, infrared light, or nuclear radiation. (See Figure 2.11.)

Bark content

Bark is an unwanted component in chips, because it has a negative influence on the pulp produced. Typical bark content specifications for debarked wood are 0.5% to 2%. Bark consists of 20% to 40% extractives, 20% to 30% cellulose, and generally more lignin than the corresponding wood. Bark fibers are also very short when compared to wood fibers. Therefore, the presence of bark increases the alkali consumption and decreases the pulp yield and pulp strength. This is illustrated in Figures 2.12 & 2.13.

The high extractive content can lead to scaling in the evaporators and also gives rise to pitch problems later in the papermaking process.

When bleaching to a specified brightness, the consumption of bleaching chemicals is higher than for bark-free wood.

The bark content is determined by hand measuring—a rather time-consuming process.

Other contaminants

Besides bark and decayed wood, chips normally contain a small fraction of non-wood contaminants such as small rocks, sand and dirt, metal, and plastics.

If the amount is too high, these contaminants can cause wear and tear to the processing equipment—especially to valves, pumps, and refiners.

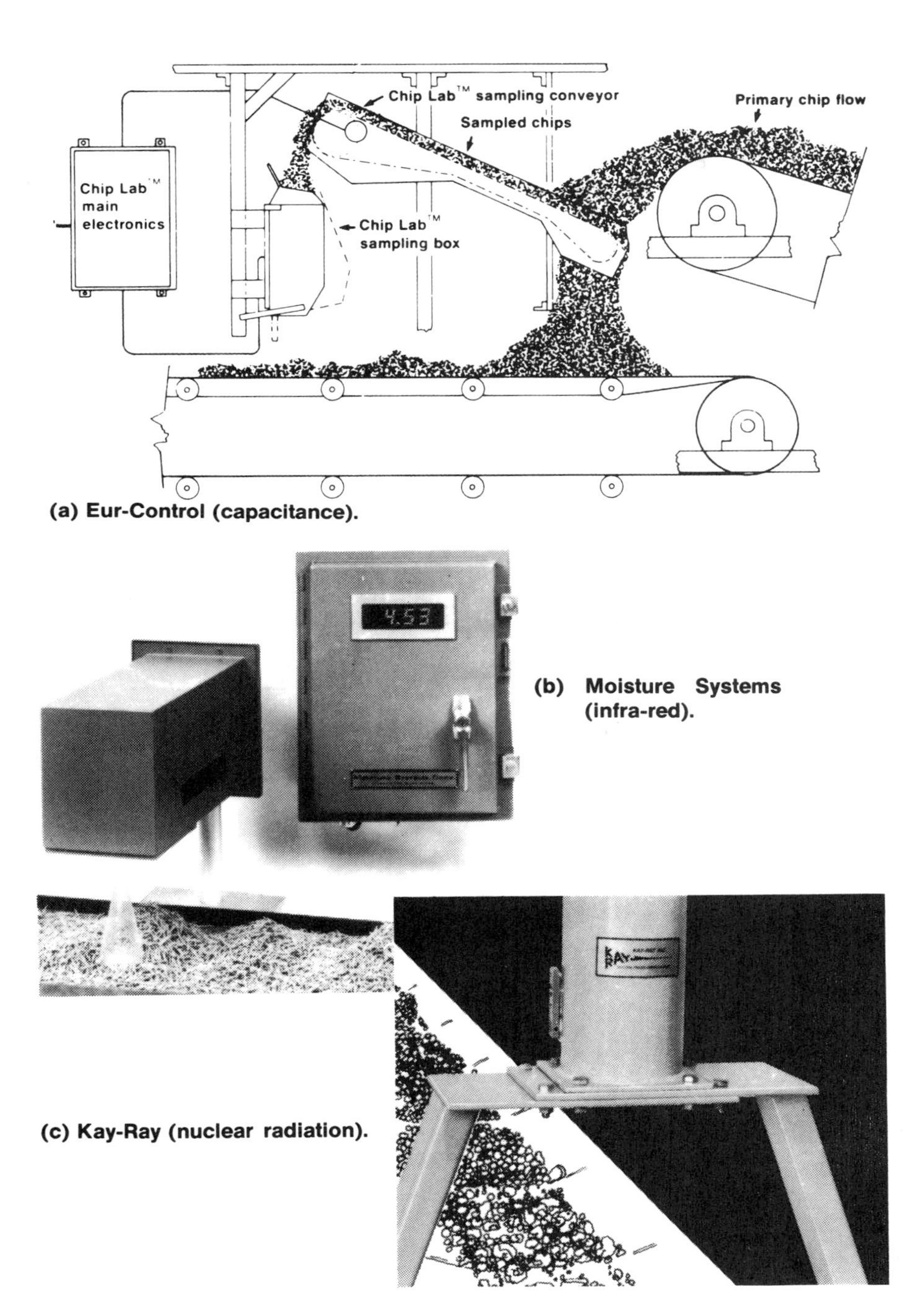

Figure 2.11 Equipment for chip moisture control.

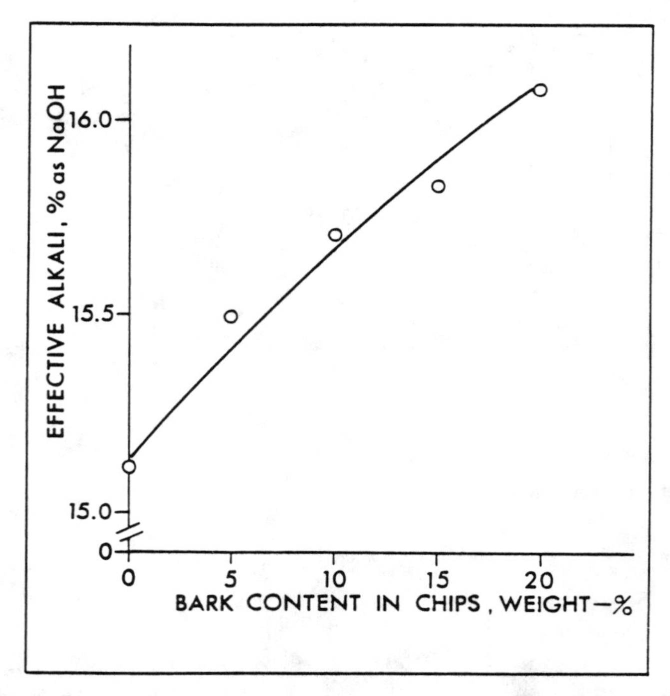

Figure 2.12 Alkali consumption vs. bark content.

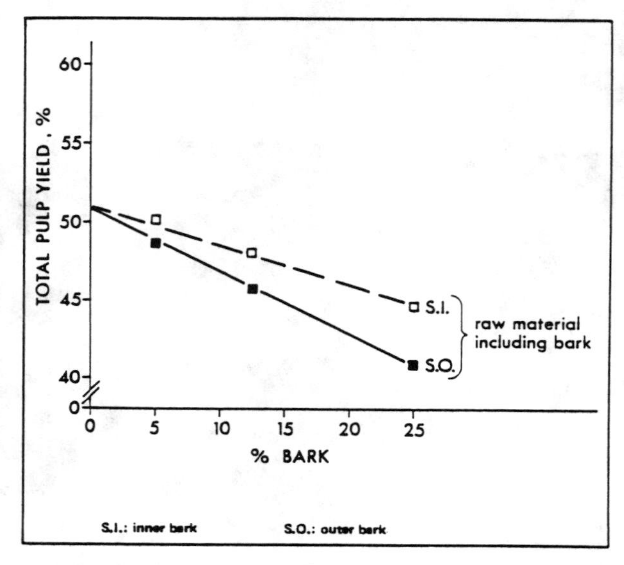

Figure 2.13 Pulp yield vs. bark content.

Questions

1. List at least five variables that will affect the chip quality.
2. Identify the main difference between softwood and hardwood fibers.
3. List the main reasons why wood quality can vary within the same wood species.
4. Identify the three main types of chips, based on the wood source.
5. List some negative effects connected with long time chip storage.
6. Give some guidelines for successful chip pile management.
7. Explain why decayed wood results in bad pulping.
8. Explain why chip thickness is the most important chip size parameter in kraft pulping.
9. Specify the optimal chip thickness for kraft pulping.
10. Give some reasons why a narrow chip size distribution is an important chip quality parameter.
11. Describe how a classifier works.
12. Explain why chip bulk density is an important chip quality parameter.
13. Describe how chip bulk density is measured.
14. Explain why the chip moisture should be known before the chips are pulped.
15. Specify the normal range for chip moisture content.
16. Explain why bark is an unwanted component in chips.
17. Name some typical contaminants in chips.

Principal Sources For Figures

Hatton, J.V., Ed. "Chip Quality Monograph," in *Pulp and Paper Technology Series No. 5.* Atlanta and Montreal: Joint Textbook Committee of the Paper Industry, 1979.

Kocurek, M.J., and Stevens, C.F.B., eds. *Pulp and Paper Manufacture.* Vol. 1: *Properties of Fibrous Raw Materials and Their Preparation for Pulping.* 3rd ed. Atlanta and Montreal: Joint Textbook Committee of the Paper Industry, 1983.

Smook, G.A. *Handbook for Pulp and Paper Technologists,* Edited by M.J. Kocurek. Atlanta and Montreal: Joint Textbook Committee of the Paper Industry, 1982.

Fig. 2.1 Kocurek, Fig. 17, p. 18.
Fig. 2.2 Kocurek, Fig. 19, pg. 19.
Fig. 2.3 Kocurek, Fig. 67, p. 67.
Fig. 2.5 Kocurek, Fig. 108c, p. 130.
Fig. 2.6 Hatton, Fig. 13, p. 286.
Fig. 2.7 Kocurek, Fig. 123, p. 150.
Fig. 2.9 Hatton, Fig. 19, p. 292.
Fig. 2.10 Hatton, Fig. 21, p. 294.
Fig. 2.11 Kocurek, Fig. 122, p. 149.
Fig. 2.12 Hatton, Fig. 15, p. 288.
Fig. 2.13 Hatton, Fig. 16, p. 289.

3
Chipping and Chip Handling

The production of uniform chips is critical to producing a high quality pulp with minimum variation in that quality. Chip thickness is the most important variable in chipping, along with the uniformity of chip sizes. Fines, pin chips, and oversize chips reduce pulp quality and may complicate digester operations.

Chipping and screening are operations discussed in terms of principles and the varieties of equipment utilized. The most common chipper is the disc type. Screen types include the gyratory screen, flat screen, disc screen, and drum screen. The objective is to produce a consistent quality of uniform chips to the digester.

In addition to fines, pin chips, and oversize chips, the presence of contaminants such as bark, dirt, metal, plastic and rock will adversely affect equipment and pulp quality.

Chipping

Objective of chipping

The objective of chipping is to reduce logs and wood product residuals into smaller pieces of relatively uniform chip size. The chips must be fairly small, or it would take too long to fully impregnate them with cooking liquor in the digester, resulting in an overcooked outside and an undercooked center.

The chips cannot be too small, however. More fines and pin chips are formed when smaller chips are chipped. Also, the average fiber length is shorter and more fiber damage is created with decreasing chip size. This ultimately leads to a weaker pulp from too small chips.

Generally, the ideal chip dimensions are a length of 10 mm to 30 mm and a thickness of 2 mm to 5 mm. For chemical pulping, chips need to be uniform in thickness and length, relatively free from damaged or short fibers, and should pack well in the digester.

Chipping equipment

The most common chipper is the disc chipper. It consists of a flat rotating disc, 2.4 m to 3 m in diameter. A number of knives (4 to 20) are mounted radially on the disc. Figure 3.1 shows a disc chipper where the hood is lifted and the chipper knives are exposed.

Pulpwood is fed from a conveyor into the feed spout of the chipper. The end of the log is struck by the knives at an angle of 35° to 40° from the longitudinal axis. Chips are then sliced off, the length of the chips being determined by the knife setting.

As the wedge-shaped knife blade cuts into the wood, pressure causes the wood to split along the grain, breaking off chips. The distance between these splits determines the thickness of the chips. (See Figure 3.2.)

This split distance depends on wood properties, the set cut length, and the wedge angle of the knife blade. Chip thickness cannot be directly controlled, but since wedge angle and length of the cut-off chip has a big impact on thickness, it can be indirectly controlled. Figure 3.3 shows how closely chip thickness and chip length are related.

Once the chips are cut, they escape through openings in the disc between the knives. The chips are then discharged either by falling through the bottom of the chipper casing or by being blown out over the top.

A chipping headrig is another type of chipping equipment that is used in sawmills. The headrig is designed to convert logs into lumber without simultaneously producing slabs or large amounts of sawdust. Instead, the outer parts of the log are made into chips. Figure 3.4 shows a chipping headrig.

The logs are fed horizontally into the chipping headrig. Rotating cutterheads, each carrying several knives, remove wood from the log in the form of single chips. Thus, the log is formed into the desired shape and then cut into boards.

The chips from chipping headrigs are normally of good quality, although they might contain a higher fraction of pinchips and have a lower bulk density than ordinary chips.

Chipping operation

It is very important to keep the chipper knives sharp at all times. Dull knives create chips with a wider size distribution containing more fines and damaged chips than normal.

Typically, knives are changed every eight hours, but if damaged by tramp metal, rocks, or other contaminants, they must be changed immediately to avoid producing chips of low quality.

The chipping operation is also affected by the kind of wood being chipped. Different chipper settings are used for different species. Frozen wood also requires specific settings. Moisture content also affects the chipping operation, and dry wood results in thinner chips containing more fines and pinchips than chips produced from green wood.

Figure 3.1 Chipper disc with knives.

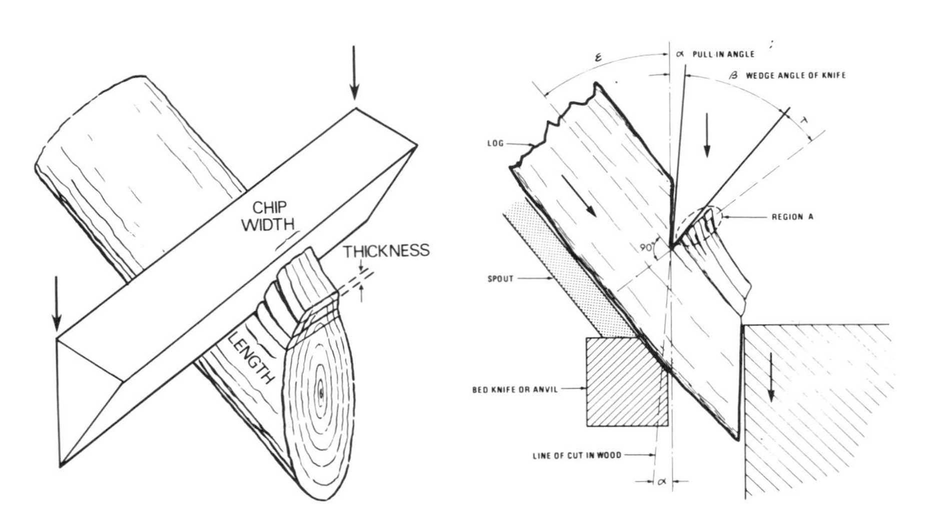

Figure 3.2 Sketch of chipping action.

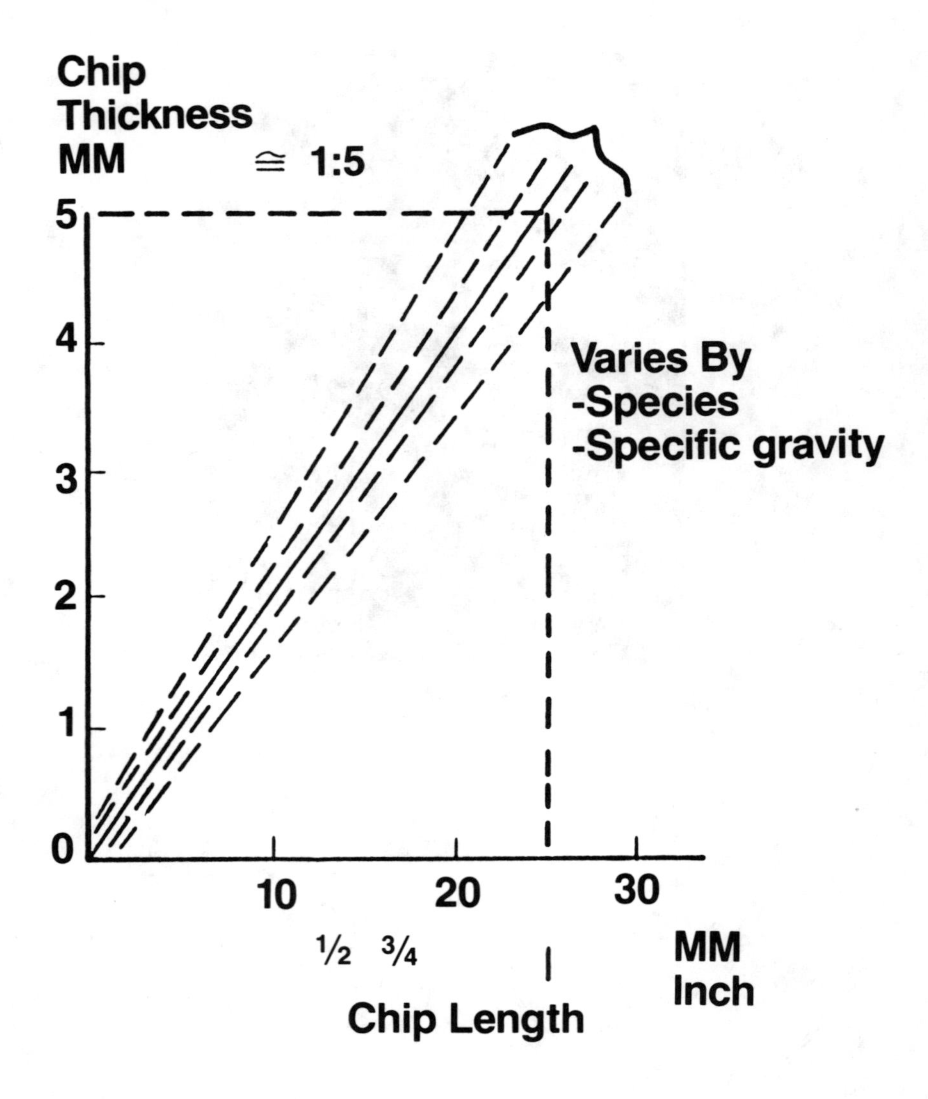

Figure 3.3 Chip thickness vs. chip length.

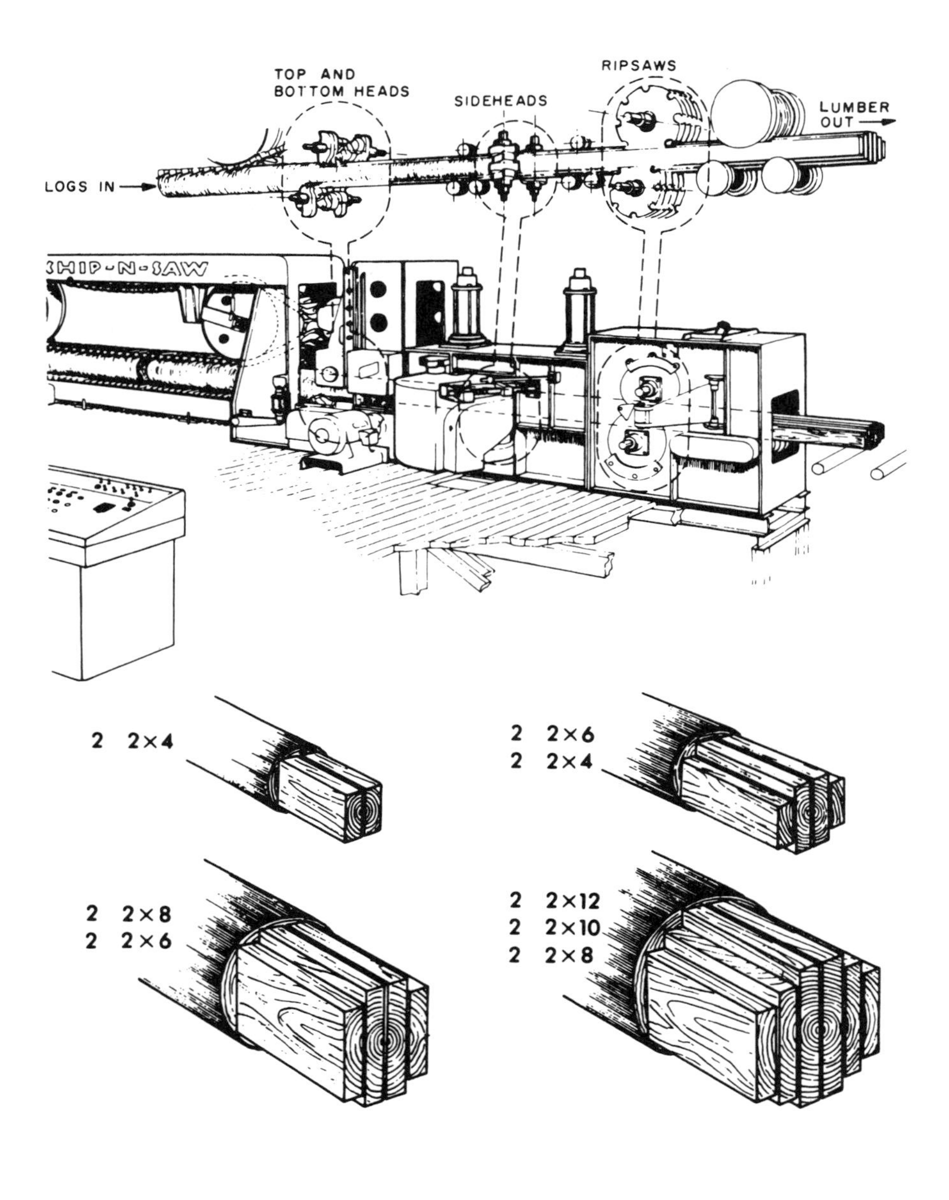

Figure 3.4 Chipping headrig and boards cut in a chipping headrig.

Chip Screening

Objective of chip screening

Chip screening removes the oversize and undersize fractions so that chips of a narrow size distribution are created that fulfill the mill's chip quality target. (See Figure 3.5 & Figure 3.6.)

Another objective of chip screening is to remove some of the small size contaminants, such as bark, sand, and grit.

Pulping is performed under conditions ideal for chips of a certain size. A wide range in chip size will result in high proportions of overcooked and undercooked chips. Big, thick chips would be undercooked and give a high percentage of knotter and screen rejects. When the pulp is screened through knotters and screens, undercooked chips and fiber bundles are separated from the acceptable fibers. Small chip fragments would be overcooked, giving a pulp of lower yield and lower strength. In kraft pulping, chip thickness is the most important size parameter. The level of pulp screenings increases dramatically with chip thickness as shown in Figure 3.7. Note that this occurs at equal target kappa numbers.

Oversize and undersize fractions can also cause mechanical problems. For example, continuous digesters are sensitive to high fractions of pinchips and fines, which can obstruct the liquor circulation screens.

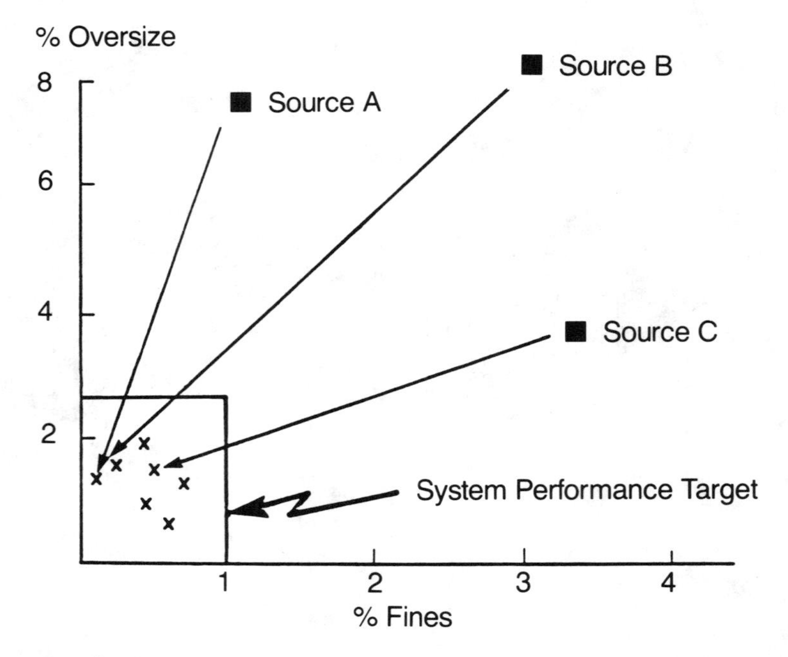

Figure 3.5 Quality target for chips.

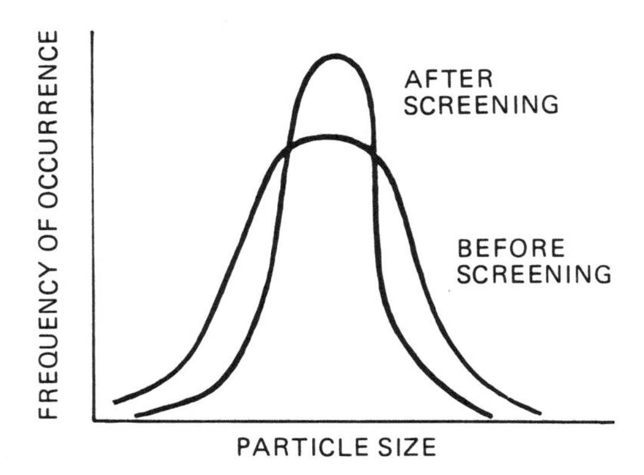

Figure 3.6 Chip size distribution before and after screening.

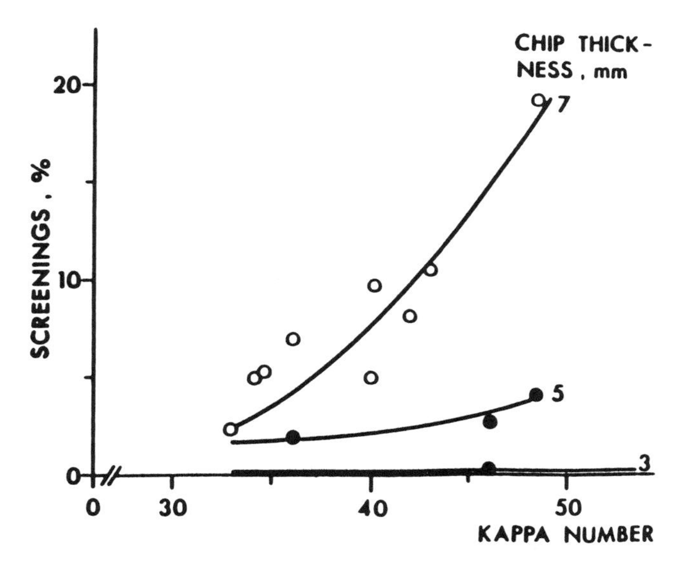

Figure 3.7 Percent pulp screen rejects vs. chip thickness.

The value of screening the chips can be summarized as follows:

- Narrower chip size distribution reduces pulp mill variability and increases productivity.
- Reduced screen rejects increase yield, decrease chemical and energy costs, and improve pulp quality.
- Dirt and bark removed with the fines fraction reduce pulp dirt, i.e., improve appearance.

Chip screening principles

Chip screening is an operation that separates material according to particle dimensions—normally size. Older chip screens are based only on size separation. There *are* new chip screen designs that separate chips based on thickness, since this parameter is proven to be the most important in kraft pulping.

Normally, chips are fed to a first screen with surface openings large enough to pass through the acceptable and undersize fractions while retaining the oversize fraction. The acceptable and undersize fractions are then screened through a second screen. This second screen retains the acceptable fraction but allows the fines and other undersize material to pass through.

A chip can pass through a screen opening only if two of its dimensions are smaller than the dimensions of the opening. It might not pass through for many reasons: wrong orientation; not enough exposure to the opening because of other chips impeding; or clumping of chips due to ice, moisture, or pitch on the chip surface.

The quality parameter used for evaluating chip screening is called *screen removal efficiency*. The chip size distribution is measured before and after screening. The removal efficiency of a size fraction is then calculated as:

$$\frac{\text{weight of fraction before—weight of fraction after}}{\text{weight of fraction before}}$$

A screen's efficiency depends on several factors. Feed composition, feed rate, screen surface area, screen openings, and screen motion.

The following is an example of how screen removal efficiency is calculated:

100 kg of unscreened chips with the following size distribution, determined in a lab classifier, were screened.

10 kg overthick and oversize
75 kg accepts
10 kg pinchips
5 kg fines

After screening, the accepts were classified again with the following results:

1.5 kg overthick and oversize
72 kg accepts
9 kg pinchips
1.0 kg fines

This gives the following results:

overthick and oversize: $\frac{10 - 1.5}{10} = 0.85$

accepts: $\frac{75 - 72}{75} = 0.04$

pinchips: $\frac{10-9}{10} = 0.10$

fines: $\frac{5-1}{5} = 0.80$

Removal efficiencies in % :

overthick and oversize	85%
accepts	4%
pinchips	10%
fines	80%

Chip screens

Flat gyratory and vibratory screens

There are several kinds of chip screen designs used in the industry today. The most common screen is the flat inclined gyratory screen. (See Figure 3.8.) The most common flat inclined gyratory screen configuration has two screen decks sloped 5° to 10° from the horizontal plane. The upper deck has openings of 3.8 cm to 5 cm, while the lower deck's openings are 6.4 mm to 9.5 mm. Normal dimensions are a width of 1.2 m to 3.7 m and a length of 2.4 m to 6 m.

Screen plates can have openings of different shapes. Usually plates perforated with round holes, or plates of wire mesh with square openings are used. Figure 3.9 shows different types of chip screen openings.

In recent years, plates with rectangular holes and slots have become common, since they separate chips according to *thickness* rather than *size*. It is possible to change plates and vary the screen openings; however, this can be a time consuming and difficult procedure.

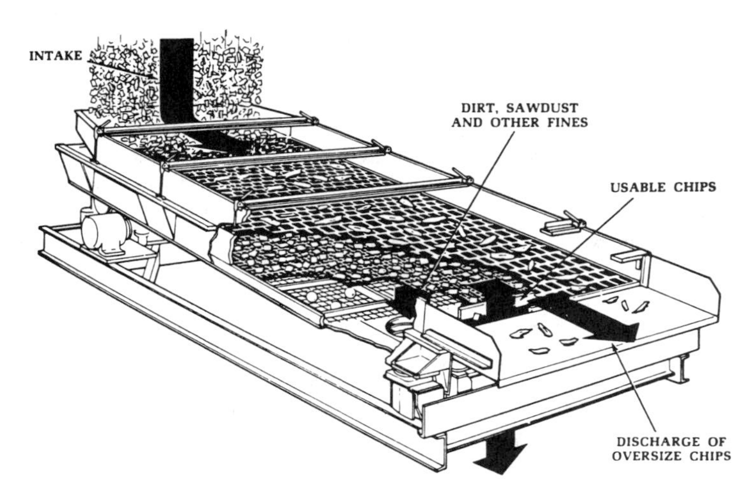

Figure 3.8 Flat inclined gyratory screen.

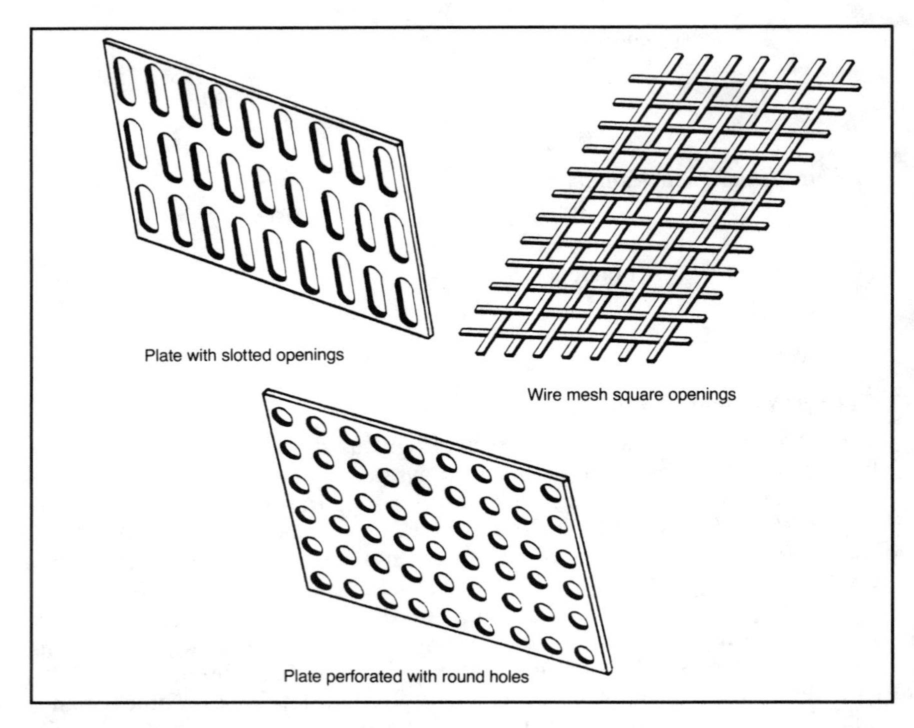

Figure 3.9 Types of chip screen openings.

The screens agitate the chips with a gyratory motion in the horizontal plane. This motion is imparted to the screens by an eccentric drive located at one end of the screens. The speed of the gyratory motion is about 200 rpm and the amplitude about 5 cm.

Vibratory screens are similar to gyratory screens in most respects except for the fact that the screens are agitated by vibration in the vertical plane. This vibration is more violent than the gyratory motion, producing a greater variety of chip orientation. Consequently, it is believed to be more effective in removing undersized material from the chips.

Drum screens

This kind of chip screen is cylindrical in design, sloping along and rotating around the axis of the cylinder. (See Figure 3.10.)

The normal size of a drum screen is a diameter of 1.2 m to 4 m and a length of 2.4 m to 8 m. The chips enter the drum at the upper end. The drum rotation makes the chips tumble, providing good mixing and changing orientation as they move through the drum towards the lower end.

In the entrance zone there is a perforated section with small openings that let undersized material pass while retaining acceptable chips and oversized

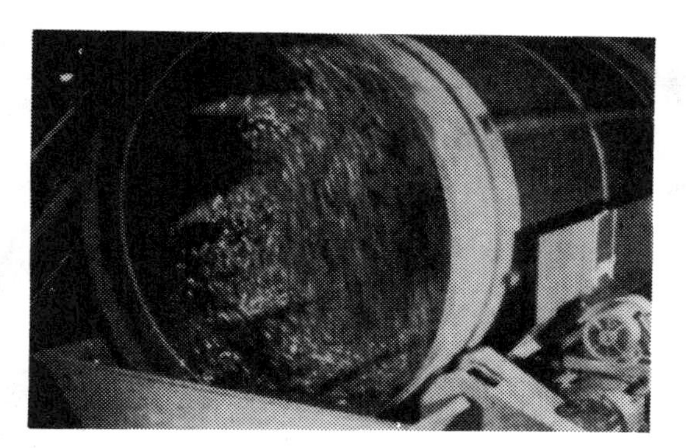

Figure 3.10 Drum screen.

material. Further down, there is a second, perforated section with openings large enough to allow acceptable chips to move through. The oversized material then exits through the lower end of the drum.

Disc screens

The newest chip screen design available is the *disc screen.* It is designed to pass or retain chips based on *thickness* only. It is composed of parallel shafts on which discs are mounted. The discs are arranged in a staggered pattern, leaving slot openings with a width of 5 mm to 13 mm between the discs. (See Figure 3.11.) All shafts rotate in the same direction. The discs are toothed, which helps to transport the chips over the screen surface. A normal size disc screen has a width of 1.8 m to 3.4 m and a length of 1.8 m to 2.7 m. A screen contains 10 to 20 shafts.

The chips are fed onto one end of the screen. The rotation of the toothed discs agitates the chips and causes each chip to upend so that one of its longer dimensions is oriented perpendicularly to the slot opening and the other one lengthwise to it. The chip can then pass through if its smallest dimension—

Figure 3.11 Disc screen.

Figure 3.12 V-disc screen.

its thickness—is less than the width of the slot. Overthick chips are rejected by not passing through slots.

Disc screens with the shafts mounted to form a "V" are called V-disc screens. A V-disc screen is shown in Figure 3.12.

Screens rejects

Fines and pinchips

Fines are normally disposed of as fuel and burned in the hog fuel boiler. They can also be recovered as raw material for sawdust pulping. Wood loss in the screening operation is generally due to the loss of fines only, since the oversize fraction normally is rechipped and recirculated to the chip flow.

If pinchips are separated in the screening operation, they are usually metered back into the chips flow to control variability. They can also be sent to a separate pulping system.

Oversize and overthick chips

The oversize and overthick fractions are normally rechipped and recycled to the chip flow. There are situations, however, when these fractions are disposed of as fuel or even as landfill.

Rechipping

Rechippers are used in almost every mill that screens its chips. Initially, a modified disc chipper or a hammermill that crushed the oversize chips to smaller fractions was used for rechipping.

A new design, the chip slicer, does a good job in reducing the size of overs.

The chip slicer is designed to slice oversize chips lengthwise, parallel to the grain, creating new, thin chips and reducing the thickness of the original overthick chip. (See Figure 3.13.)

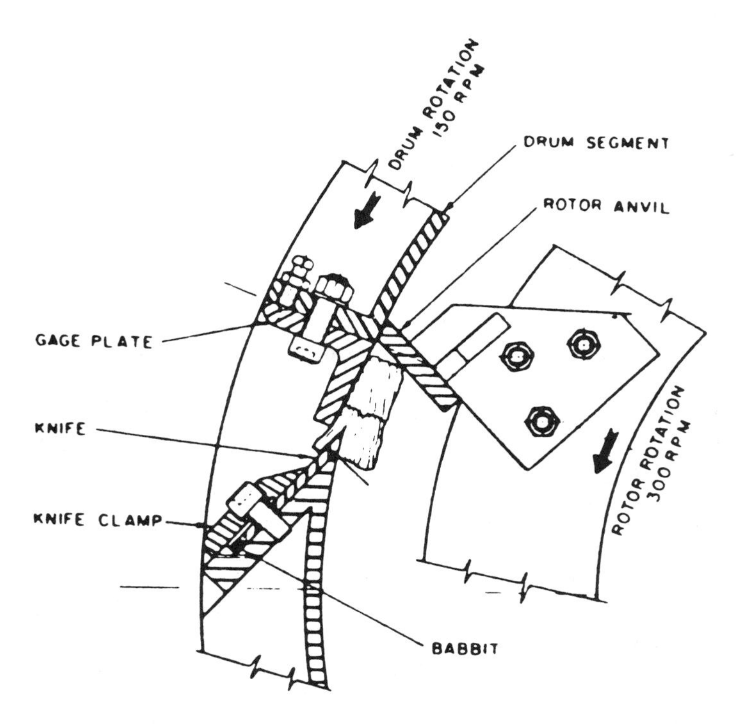

Figure 3.13 Chip slicer.

This method has some real benefits. First, it causes little damage to the fibers when compared to the old methods. It also produces less fines, which means that the slicer discharge can be recycled directly to the screened chip flow without rescreening. Discharges from disc chippers and hammermills contain a large amount of fines and are normally recycled to the chip screen. This causes an extra screen load and can reduce total screening efficiency. Also, by using a chip slicer for all oversize and overthick chips, one has a great deal of control over chip thickness.

Chip cleaning

Objective of chip cleaning

The objective of chip cleaning is to remove contaminants from the chips that would otherwise reduce pulp quality or cause wear and damage to the equipment in the chips and pulp flow. Even though contaminants usually compose a small percentage of total chips weight, they can cause serious problems if not removed.

Contaminants

The different kinds of contaminants found in chips are:

- Rocks
- Tramp metal
- Plastics
- Sand and grit.

Large rocks and metal can damage equipment on continuous digesters such as pressure feeders. Refiners will also be damaged if a rock or a chunk of metal gets between the refiner plates.

Sand and grit cause wear on all kinds of equipment through their abrasive action.

Rocks, sand, and grit are all contaminants connected with the wood source. These contaminants follow the logs through debarking and chipping, some remaining with the chips.

Metal can enter the system with the chips in the form of rusty nails or barbed wire. Metals can also come from the process equipment in the form of bolts or broken bearings.

Plastics are a serious contaminant that are difficult to remove once in the chips. Currently there are no effective removal techniques available for plastics. Plastics reduce pulp quality by decreasing appearance. They will not dissolve during cooking, but will instead form small particles that show up as specks in the pulp. Almost all plastic enters the chip flow at the mill site. The most common sources are different forms of packaging such as food wrappers and packaging for equipment parts. Broken plastic parts and plastic strapping are also commonly found. The only way to avoid product quality problems associated with plastics is *education*—making people aware of what happens when they throw plastic products into the chips.

Equipment

If the chips are screened, the contaminants that are oversize, such as large rocks and chunks of metal, will be removed with the oversize fraction. Most undersize contaminants will be removed with the fines. However, the contaminant fraction having the same size range as the accepted chips must be removed from the chips by taking advantage of any difference in physical

Figure 3.14 Magnet.

properties between them and the chips. The methods that do this are all referred to as chips cleaning.

Magnets

A magnet will remove ferrous metals, such as iron, from the chip flow. It will not remove metals such as copper, aluminum, or lead, however. The magnet is normally mounted above a conveyor belt carrying chips. The magnet must be strong and positioned so it can pull metal objects through the chips when the belt is moving at full speed and the chips layer on the belt is of maximum thickness. (See Figure 3.14.) The magnet must be self-cleaning or have a routine cleaning schedule that assures the removal efficiency is kept at design level.

Chip washers

In a chip washer the chips are passed through a water bath where they have a certain retention time to allow contaminants to separate after chips and water have been mixed. Basically, a chip washer will remove contaminants that sink in water: stones, metal, sand, and grit. A long enough retention time must be provided to effectively separate fine sand and grit since they

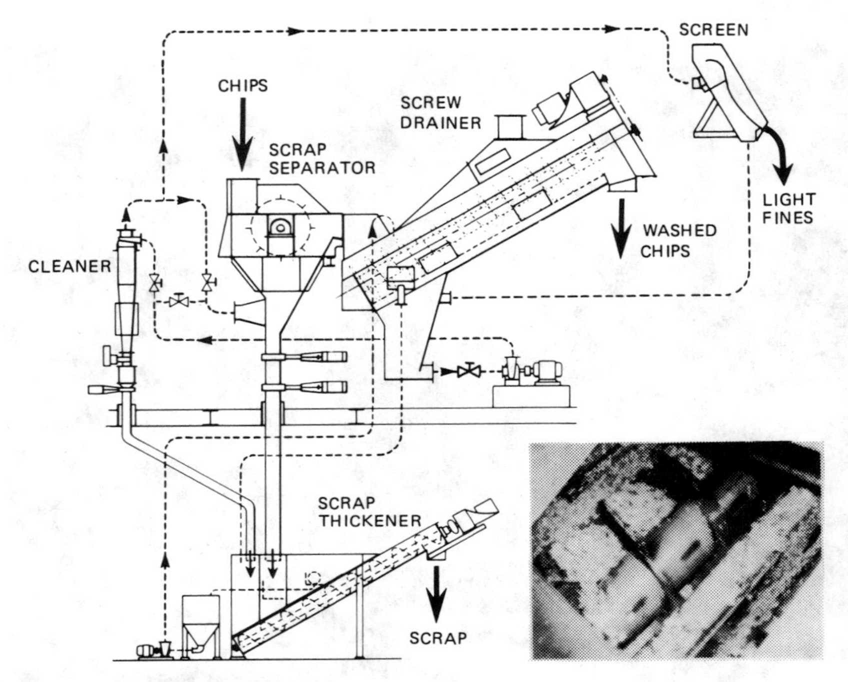

Figure 3.15 Chip washer.

tend to stick to the rough surfaces of the chips. They also have slower settling rates; that is, they sink more slowly.

Some bark and fines can also be removed by a chip washer. Since they do not sink, they will not settle out of the wash water and must be removed by filtering the wash water.

Chips will absorb water in the washing process. For example, if they contain 50% moisture before washing, they may contain 53% to 58% moisture as they leave the washer. The increase in moisture content depends on species, retention time in the washer, and chip size distribution.

Pneumatic cleaners

A pneumatic cleaner separates chips from denser material by letting an air flow carry away the lighter chips while heavier particles not carried away are collected as rejects.

A pneumatic cleaner is most efficient in removing larger objects such as rocks and pieces of metal. It will also remove fragments of knot wood that can have a specific gravity much higher than that of normal wood. The air flow and the material flow in a pneumatic cleaner must be controlled and balanced so that the removal efficiency is kept high while the wood loss is kept low.

Fine particles of wood, sand, and grit can also be removed in a cleaning system based on air. In a pneumatic cyclone chip cleaner, chips enter at the top and, due to gravity, they drop down and are discharged through an airlock at the bottom. The fan creates a suction pressure and small, light particles that can pass through the screen are carried away with the air through the fan. Figure 3.16 shows a pneumatic chip cleaning system including a cyclone. Figure 3.17 shows a close-up of the cyclone.

Normally the chips are cleaned after screening. When the oversize fraction from screening goes on to a rechipper or chip slicer, rocks and metal parts must be removed first to avoid damaging the equipment. This is usually done in some kind of pneumatic cleaner.

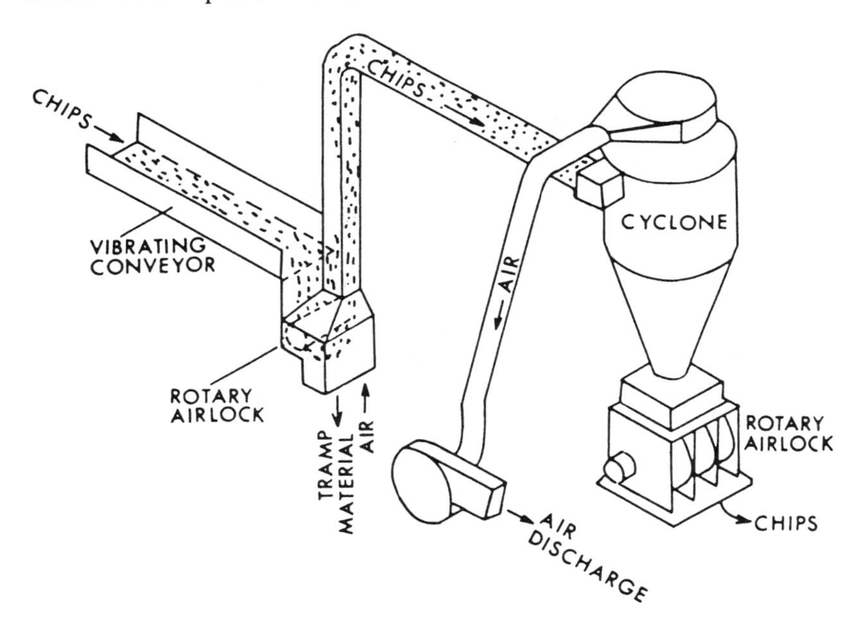

Figure 3.16 Pneumatic chip cleaning system.

Whole tree chips

Whole tree chips contain a high percentage of foliage, bark, and dirt compared to roundwood or residual chips. Much of the bark is still attached to the chips and is not removed by conventional screening and washing.

In order to upgrade whole tree chips to a quality level resembling roundwood and residual chips, the chips would have to go through a more extensive cleaning treatment than roundwood and residual chips. One such method has three main steps:

1. *Conditioning.* The chips are either stored in piles for several weeks or steamed for several minutes in order to weaken the bark-to-wood bond.
2. *Agitation in a pulper.* Chips and water are added to a pulper where bark

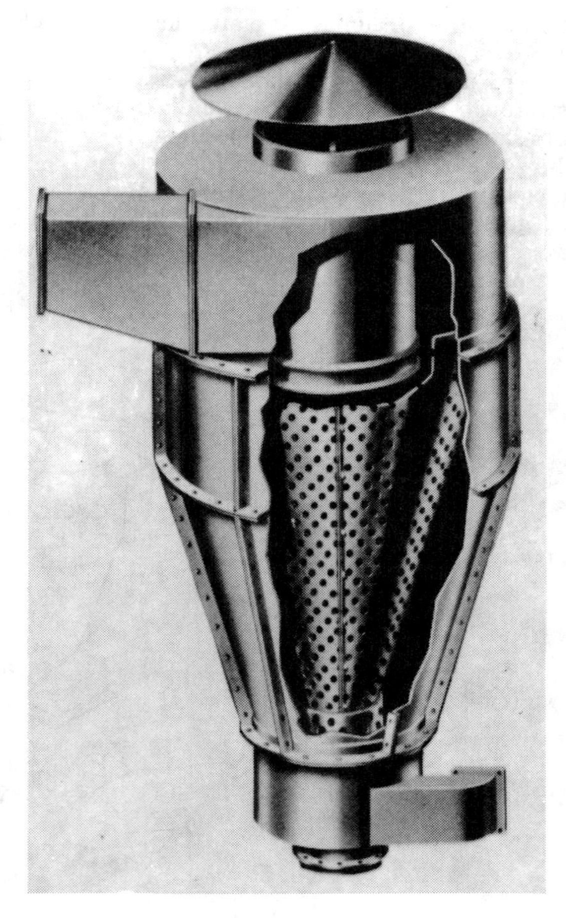

Figure 3.17 Pneumatic cyclone chip cleaner.

and foliage are detached and broken down into small fragments by the action of an impeller.

3. *Washing.* Bark and foliage fragments are separated from the chips by removing them with the wash water through a screen at the bottom of the pulper. The fragments are then separated from the wash water, which is recycled.

Questions

1. Describe how a disc chipper operates.
2. Explain the importance of keeping chipper knives sharp at all times.
3. State the purpose of chip screening.
4. List some improvements in the pulping operation and pulp quality that can be achieved by screening chips.
5. Discuss the principles of chip screening.
6. Give the definition of the term "screen removal efficiency."
7. Calculate screen removal efficiencies.
8. Recognize different types of chip screen designs.
9. Explain how different types of chip screens operate.
10. Explain what advantage disc screens have over other screen designs.
11. Discuss different ways of handling screen rejects.
12. Identify the objective of chip cleaning.
13. List some typical contaminants in chips.
14. Describe the operating principles of a magnet, a chip washer, and a pneumatic chip cleaner.
15. Explain why full-tree chips need more extensive cleaning than other chips.
16. Specify the major steps in a chip cleaning system for whole tree chips.

Principal Sources For Figures

Hatton, J.V., Ed. "Chip Quality Monograph," in *Pulp and Paper Technology Series No. 5.* Atlanta and Montreal: Joint Textbook Committee of the Paper Industry, 1979.

Kocurek, M.J., and Stevens, C.F.B., eds. *Pulp and Paper Manufacture.* Vol. 1: *Properties of Fibrous Raw Materials and Their Preparation for Pulping.* 3rd. ed. Atlanta and Montreal: Joint Textbook Committee of the Paper Industry, 1983.

Smook, G.A. *Handbook for Pulp and Paper Technologists,* Edited by M.J. Kocurek. Atlanta and Montreal: Joint Textbook Committee of the Paper Industry, 1982.

TAPPI, *Introduction to Pulping Technology,* TAPPI Home Study Course, no. 2. Atlanta: TAPPI Press, 1976.

TAPPI, *Notes from Chip Preparation and Quality Seminar.* Washington, D.C., 1987.

Fig. 3.1 Kocurek, Fig. 109a, p. 131.
Fig. 3.2 Kocurek, Fig. 108, p. 130.
Fig. 3.3 TAPPI, Chip Preparation and Quality Seminar, Session 6..
Fig. 3.4 Kocurek, Fig. 112, p. 136.
Fig. 3.5 TAPPI, Chip Preparation and Quality Seminar Session 6.
Fig. 3.6 TAPPI, Chip Preparation and Quality Seminar Session 6.
Fig. 3.7 TAPPI, Chip Preparation and Quality Seminar Session 6.
Fig. 3.8 TAPPI, *Introduction to Pulping Technology,* Fig. 4.11, p. IV-7.

Fig. 3.9 Hatyton, Fig. 1.3, p. 76–77.
Fig. 3.10 Kocurek, Fig. 114d, p. 137.
Fig. 3.11 Kocurek, Fig. 115c, p. 139.
Fig. 3.12 Kocurek, Fig. 115b, p. 139.
Fig. 3.13 Smook, Fig. 3.30, p. 30.
Fig. 3.14 Kocurek, Fig. 116, p. 142.
Fig. 3.15 Kocurek, Fig. 118, p. 144.
Fig. 3.16 Kocurek, Fig. 116b, p. 143.
Fig. 3.17 Kocurek, Fig. 116a, p. 143.

4

Kraft Pulping: Chemistry and Process

In kraft pulping the active chemicals consist of sodium hydroxide (NaOH) and sodium sulfide (Na_2S). The primary dead-load chemical also present is sodium carbonate (Na_2CO_3). The change of these chemicals expressed as the weight of chemical per weight of wood, along with liquor to wood ratio, chemical concentration, and residual chemical constitute key liquor variables. Time and temperature are the key operating variables.

Chemical reactions occur between the active alkali and effective alkali (actual amount of caustic OH present), and the wood components. Most of the chemical is consumed by carbohydrates, and pulp strength is determined by the degree of cellulose and hemicellulose removal. The presence of sodium sulfide, Na_2S, produces hydrosulfide ions, HS^-, which accelerate lignin removal. This yields a stronger pulp. Cooking is controlled by a target kappa number which measures the amount of residual lignin present at the end of the cook.

The operations in a batch digester include chip and liquor filling, impregnation of chips by liquor, heating, cooking, and blowing or emptying the digester. Each of these steps, along with their equivalents in continuous digesters, affects pulp quality and uniformity.

Chemistry of Kraft Pulping

Objective of kraft pulping

The objective of kraft pulping is to chemically separate the fibers in wood and dissolve most of the lignin contained in the fiber walls. Fiber separation is achieved by dissolving the lignin in the middle lamella that holds the fibers together. The chemicals in the cooking liquor also penetrate the fiber walls and dissolve the lignin there.

Cooking Liquor

Composition

Cooking liquor, or white liquor as it is also called, is an aqueous solution of sodium hydroxide (NaOH) and sodium sulfide (Na_2S). Mill white liquor also contains inactive "dead-load" chemicals, the primary one being sodium carbonate (Na_2CO_3).

Active components

The active components in the cooking liquor are the hydroxyl ion (OH^-) and the hydrosulfide ion (SH^-). They originate from NaOH and Na_2S, as follows:

$$NaOH \rightarrow Na^+ + OH^-$$

$$Na_2S \rightarrow 2\ Na^+ + S^{2-}$$

$$S^{2-} + H_2O \rightarrow SH^- + OH^-$$

The sulfide ion originating from the sodium sulfide reacts with a molecule of water. The result is one hydrosulfide ion and one hydroxyl ion.

The concentration and total charge of the SH^- and OH^- ions are the key elements in all reactions that take place during pulping, both in lignin dissolution and in unwanted reactions such as cellulose degradation. The total OH^- present from the original caustic and part of the original sulfide is called the effective alkali.

Reactions during digesting

Reactions with lignin

The reactions of lignin during kraft pulping are complex and not completely understood. It is known that the presence of hydrosulfide ions accelerates lignin dissolution without increasing the dissolution of cellulose.

The overall effect of all the reactions between lignin, hydrosulfide, and hydroxyl ions is that the lignin polymer is broken down into smaller molecules. These small molecules are no longer able to function as "cement" and remain in the wood structure. They are instead dissolved in the cooking liquor and leave the wood fibers.

Reactions with carbohydrates

Ideally, only the lignin would dissolve during pulping, but that is not the case. Both cellulose and hemicellulose react with hydroxyl ions during cooking. These reactions are unwanted, since they degrade the carbohydrates to smaller, soluble molecules, which lowers the yield.

More than 20% of the original wood is lost due to loss of cellulose and hemicellulose. Most of this loss takes place early in the cook. Hemicellulose is degraded faster and to a larger extent than cellulose. This is because hemicellulose consists of smaller, more branched molecules. Hemicellulose is also exposed to a higher degree in the wood structure.

Over 2/3 of all of the effective alkali changed is consumed by the carbohydrates. In order to be certain there is sufficient alkali to remove liquor, there must be a sufficient residual alkali present at the end of the cook. Modified continuous cooking (MCC) adds a portion of the alkali at the beginning of the cook and the rest at a later part. This reduces reactions with the carbohydrates.

One reaction that takes place during cooking is called *peeling*. It affects both cellulose and hemicellulose. In the peeling reaction, the sugar units at one end of the cellulose or hemicellulose chain are removed, one by one. (See Figure 4.1.)

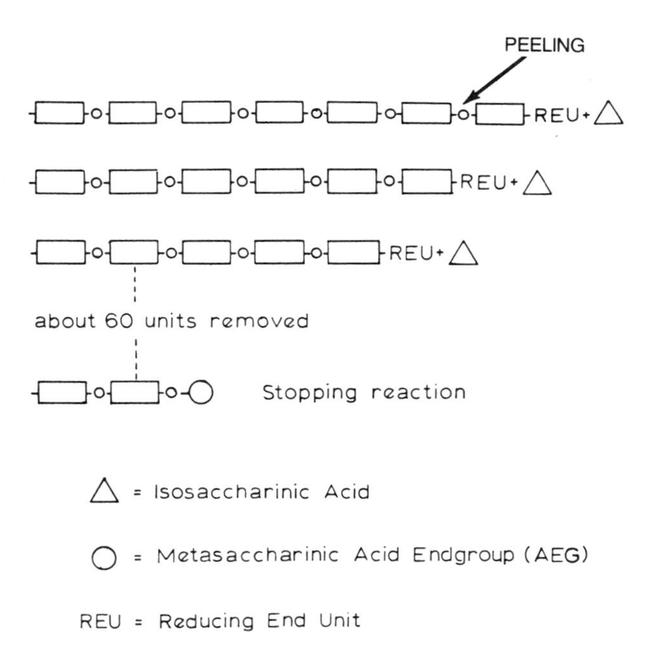

Figure 4.1 Stepwise depolymerization of carbohydrates in alkaline cooking liquor (peeling reaction).

Another reaction is called *stopping*. During this reaction the end of the chain is stabilized so no more peeling reactions can take place.

A reaction that becomes important at temperatures above approximately 170°C is *alkaline hydrolysis*. In this reaction the cellulose chain is cleaved into two parts, creating a new end group that can undergo peeling. This is not good, since more cellulose will be dissolved, and the pulp yield will decrease.

Reactions with extractives

Extractives react with and consume cooking chemicals as well. Most of the extractives are dissolved during cooking. Some of the dissolved extractives can be recovered as by-products from the kraft process. Tall oil and turpentine are such products.

Some substances belonging to the extractives group are very difficult to dissolve and remain in the pulp. They are referred to as "non-saponifiables."

Pitch deposits on equipment is a serious problem for many pulp and paper mills. Pitch consists of small aggregates of undissolved extractives.

Typical pulp yields of wood components

This table shows how much of each wood component can be expected to remain after pulping when pine is pulped to a 50% yield.

	Original Wood (Pine)	50% Yield Pulp
Lignin	30 kg	5 kg
Cellulose	45 kg	40 kg
Hemicellulose	20 kg	5 kg
Extractive	5 kg	0
	100 kg	50 kg

The Kraft Pulping Process

Standard kraft pulping terms

Introduction

In North America the practice is to express all sodium components on the basis of the equivalent amount of sodium oxide (Na_2O). In Europe it is more common to select sodium hydroxide (NaOH) as the basis.

The concentration unit used for reporting laboratory work is usually grams per liter. Pounds per cubic foot is still the basis for many mill reports, however.

The use of sodium oxide, Na_2O, as a standard may seem odd since sodium oxide never appears in the pulping solution. The terminology has developed over many years, though, and is totally accepted.

Definition of terms

TAPPI (United States) Definitions:

The following definitions are from TAPPI Technical Information Sheet 0601-05.

1. *Total chemical:* All sodium salts, expressed as Na_2O.
2. *Total alkali:* $NaOH + Na_2S + Na_2CO_3 + 1/2\ Na_2SO_3$, all expressed as Na_2O.
3. *Active alkali:* $NaOH + Na_2S$, expressed as Na_2O.

4. *Effective alkali:* $NaOH$ + 1/2 Na_2S, expressed as Na_2O.
5. *Activity:* The percentage ratio of active alkali to total alkali.
6. *Causticizing efficiency:* In white liquor, the percentage ratio of $NaOH$ to $NaOH + Na_2CO_3$, both items being expressed as Na_2O, and being corrected for $NaOH$ content of the original green liquor in order to represent only the $NaOH$ produced in the actual causticizing reaction.
7. *Causticity:* The percentage ratio of $NaOH$, expressed as Na_2O, to active alkali.
8. *Sulfidity:* The percentage ratio of Na_2S, expressed as Na_2O, to active alkali.
9. *Reduction:* In green liquor, the percentage ratio of Na_2S to Na_2SO_4 + Na_2S + any other sulfur compounds, all expressed as Na_2O.
10. *Unreduced salt cake:* Na_2SO_4 in the green liquor, expressed as Na_2SO_4.
11. *Makeup chemical consumption:* Pounds of Na_2SO_4, or other sodium compounds expressed as Na_2SO_4, added as new chemical per ton of air dry pulp produced.
12. *Chemical recovery:* The percentage ratio of total chemical to the digesters, less the total chemical in new chemical, to the total chemical to the digesters (after correcting for any change in the liquor inventory).

NOTE: These are the U.S. definitions. Canada and Europe have a few different definitions, and use $NaOH$ as a basis, rather than Na_2O.

Analysis of Kraft Liquors

Quantitative knowledge concerning the composition of the various liquors found in the kraft process is vital for proper control. Because of the many compounds present, analytical procedures are somewhat complex. Many of the methods are based on traditional wet chemistry. However, a slow conversion to the more modern analytical techniques, such as ion chromatography, is occurring. The reader is urged not to skip this section as it provides valuable information on liquor chemistry in addition to discussing the purely analytical aspects of the subject. Following the procedure for each determination is a discussion of the method and mention of alternate ways of performing the analysis.

White and green liquors

(a) Determination of sodium hydroxide, sodium sulfide, and sodium carbonate by ABC test

Test A

Pipet a 5 mL portion of the clear liquor into a 250 mL Erlenmeyer flask. Add about 50 mL of water and 25 mL of 10% barium chloride solution. Add a few drops of phenolphthalein indicator and titrate the solution with $0.5\underline{N}$ HCl until the pink color disappears (pH 8.3). Record buret reading A. Do not refill the buret.

Test B

Add 5 mL of 40% formaldehyde to the solution after the phenolphthalein endpoint in Test A. The pink color will return, and after 30 s continue the titration until the pink color again disappears. Record buret reading B. Do not refill the buret.

Test C

Add a few drops of methyl orange to the solution after the second phenolphthalein endpoint in Test B. Continue the titration until the first trace of red appears (pH 4.0). Record buret reading C.

In test A, barium chloride is used to precipitate the carbonate in the liquor, and the sodium hydroxide and one half the sodium sulfide are titrated at pH 8.3. See curves 1 and 2 on Fig. 4.2. Curve 3 shows that one half of the carbonate would also have been titrated if it had not been precipitated. Addition of formaldehyde in test B complexes the hydrosulfide, releasing the equivalent sodium hydroxide, as shown by curve 3 on Fig. 4.3. Thus the second phenolphthalein endpoint represents the sodium hydroxide and all of the sodium sulfied, i.e., the active alkali, in the white or green liquor. Further addition of standard acid to pH 4.0 dissolves and reacts with the barium carbonate. Thus test C represents the total alkali, and the difference between test C and test B gives the sodium carbonate.

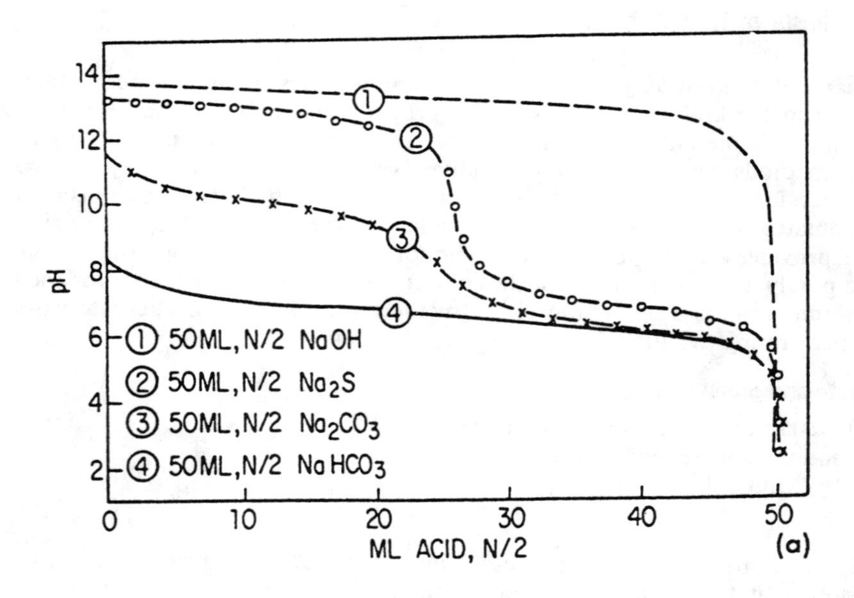

Figure 4.2 Titration curves for liquor analysis.

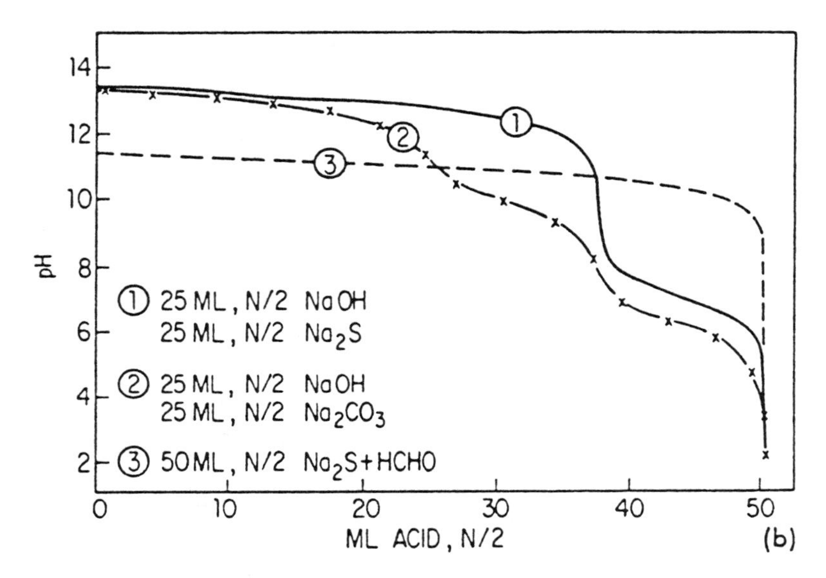

Figure 4.3 Titration cures for liquor analysis.

Conversions of ABC test results into amounts of standard acid used to titrate each of the principal alkaline species in white or green liquors are summarized below. Test results are normally expressed as g/L Na_2O.

Na_2S = 2(B-A)
NaOH = B-2(B-A) = 2A-B
Na_2CO_3 = C-B
Active alkali = B
Total alkali = C
% Sulfidity = 2(B-A)/B × 100

Calculations

These are the molecular weights for chemicals present in cooking liquor:

Chemical	Liquor	Molecular Weight (g/mole)*
sodium oxide	Na_2O	62.0
sodium hydroxide	NaOH	40.0
sodium sulfide	Na_2S	78.1
sodium hydrosulfide	NaSH	56.1
sodium carbonate	Na_2CO_3	106.0
sodium sulphate	Na_2SO_4	142.1
sodium thiosulfate	$Na_2S_2O_3$	158.2
sodium sulfite	Na_2SO_3	126.1

*Refer to Appendix A.1.2 for the definition of mole.

Assume that a lab provides the following white liquor report showing the concentrations of the different chemicals:

	As Actual Chemical	As Na_2O
NaOH	100 g/l	77.5 g/l
Na_2S	30 g/l	23.8 g/l
Na_2CO_3	25 g/l	14.6 g/l
Na_2SO_4	10 g/l	4.36 g/l
Na_2SO_3	2 g/l	0.98 g/l

Calculate total alkali, active alkali, effective alkali, and sulfidity of the white liquor.

Traditionally, laboratory results are reported as g/l of chemical, as Na_2O. Therefore,

$$\text{Total alkali} = 77.5 + 23.81 + 14.62 + \frac{0.98}{2} = 116.42 \text{ g/l as } Na_2O$$

$$\text{Active alkali} = 77.5 + 23.81 = 101.31 \text{ g/l as } Na_2O$$

$$\text{Effective alkali} = 77.5 + \frac{23.81}{2} = 89.41 \text{ g/l as } Na_2O$$

$$\text{Sulfidity} = \frac{23.81}{101.34} = 0.235 \text{ ; } 23.5\%$$

The following is a graphical representation of the relationship between active alkali, sulfidity, and effective alkali:

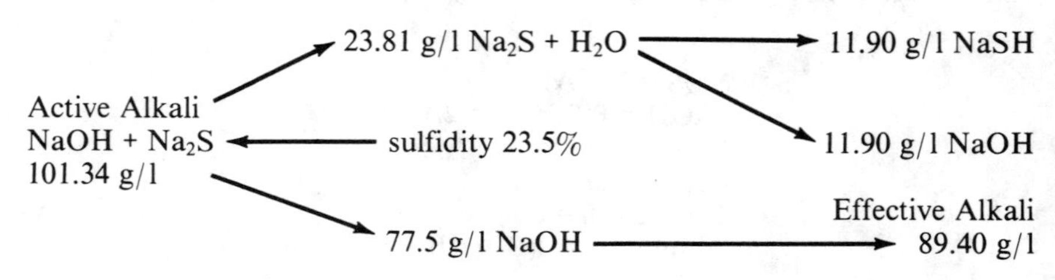

What is the relationship between g/l of chemicals as Na_2O and g/l of actual chemical? If the results were reported in g/l of actual chemical, how do we convert to an Na_2O basis? This is done in the following manner:

1. How many moles Na^+ does 100 g/l NaOH correspond to?

   One liter contains 100g NaOH. 100g NaOH is equivalent to (100g/40g/mole) = 2.5 moles NaOH; i.e., 2.5 moles Na^+.

2. How many moles Na_2O does 2.5 moles Na^+ correspond to?

   Since Na_2O contains two Na^+ ions but NaOH contains only one Na^+ ion, we need to divide by two to obtain the number of moles of Na_2O that can be formed from 2.5 moles Na^+.

   $$2.5/2 = 1.25 \text{ moles } Na_2O$$

3. How much does 1.25 moles Na_2O weigh?

$$1.25 \text{ moles} \times 62\text{g/mole} = 77.5 \text{ grams}$$

Therefore, 100 g/l NaOH is equivalent to 77.5 g/l Na_2O.

This is a summary of the calculations:

$$\frac{100\text{g NaOH}}{} \; \frac{1 \text{ mole NaOH}}{40\text{g NaOH}} \; \frac{1 \text{ mole } Na_2O}{2 \text{ moles NaOH}} \; \frac{62\text{g } Na_2O}{1 \text{ mole } Na_2O} = 77.5\text{g } Na_2O$$

For sodium sulfide, Na_2S (and the other chemicals), we do not need to divide by two since one mole of Na_2S and one mole of Na_2O contain the same amount of Na^+ ions.

Therefore, the equivalent concentrations of the other chemicals are obtained as follows:

Na_2S:

$$\frac{30\text{g } Na_2S}{} \; \frac{1 \text{ mole } Na_2S}{78.1\text{g } Na_2S} \; \frac{1 \text{ mole } Na_2O}{1 \text{ mole } Na_2S} \; \frac{62\text{g } Na_2O}{1 \text{ mole } Na_2O} = 23.81\text{g } Na_2O$$

Na_2CO_3:

$$\frac{25\text{g } Na_2CO_3}{} \; \frac{1 \text{ mole } Na_2CO_3}{106.0\text{g } Na_2CO_3} \; \frac{1 \text{ mole } Na_2O}{1 \text{ mole } Na_2CO_3} \; \frac{62\text{g } Na_2O}{1 \text{ mole } Na_2O} = 14.62\text{g } Na_2O$$

Na_2SO_4:

$$\frac{10\text{g } Na_2SO_4}{} \; \frac{1 \text{ mole } Na_2SO_4}{142.1\text{g } Na_2SO_4} \; \frac{1 \text{ mole } Na_2O}{1 \text{ mole } Na_2SO_4} \; \frac{62\text{g } Na_2O}{1 \text{ mole } Na_2O} = 4.36\text{g } Na_2O$$

Na_2SO_3:

$$\frac{2\text{g } Na_2SO_3}{} \; \frac{1 \text{ mole } Na_2SO_3}{126.1\text{g } Na_2SO_3} \; \frac{1 \text{ mole } Na_2O}{1 \text{ mole } Na_2SO_3} \; \frac{62\text{g } Na_2O}{1 \text{ mole } Na_2O} = 0.98\text{g } Na_2O$$

These calculations were performed to illustrate the actual chemical relationships between liquor components, and helps explain the rationale for using Na_2O as a common base. Often the molecular weights are expressed as a factor since they do not change. To obtain g/l as Na_2O, multiply the concentration of the chemicals with the corresponding factors.

Chemical Formula	Factor to convert weight of chemical to Na_2O equivalent
NaOH	0.775
Na_2S	0.794
Na_2CO_3	0.585
Na_2SO_4	0.436
Na_2SO_3	0.492

To calculate the rest, use the definition of each term. Having a common base of Na_2O allows us to add individual values to determine active alkali, effective alkali, etc.

The cooking liquor

Composition of typical white liquors

The following contains the chemical concentration ranges normally found in mill white liquors:

Chemical	Concentration Range (g/l as Na_2O)
$NaOH$	81 to 120
Na_2S	30 to 40
Na_2CO_3	11 to 44
Na_2SO	2.0 to 6.9
Na_2SO_4	4.4 to 18
$Na_2S_2O_3$	4.0 to 8.9

Small amounts of NaCl, potassium salts, silicate, and calcium are also present.

The only active chemicals in the pulping reactions are NaOH (sodium hydroxide) and Na_2S (sodium sulfide). The other chemicals have no direct effect on the pulping and are, therefore, called dead-load chemicals.

Sodium sulfate (Na_2SO_4) results mainly from incomplete reduction in the recovery furnace. Sodium carbonate (Na_2CO_3) is present because the causticizing is incomplete and sodium thiosulfate ($Na_2S_2O_3$) originates from air oxidation of sulfide.

Even if the dead-load chemicals do not take part in the pulping reactions, high levels of them in the white liquor are undesirable. They can cause scaling in the digester and especially in the evaporators. Also, they increase the load on the recovery furnace.

Liquor-to-wood ratio

Liquor-to-wood ratio is a term that describes the amount of total liquor per amount of dry wood in a digester. The moisture content in the chips should be included in the total liquor. Typical liquor-to-wood ratios range between three and five in mills.

Example: a batch digester is loaded with the following:

100 tons chips
50% moisture content
70 tons white liquor
80 tons black liquor

What is the liquor-to-wood ratio?
Amount of total liquor:
$100 \times 0.50 + 80 + 70 = 200$ tons
Amount of dry wood:
$100 \times 0.5 = 50$ tons

→ Liquor-to-wood ratio = 200/50 = 4.0

Alkali charge

Both effective and active alkali can be expressed as a percentage of dry-wood weight. They are then referred to as effective or active alkali charge. Effective alkali charge can range from 10% to 15% on oven-dry wood.

Example: Calculate the effective alkali charge for the following digester load:

50 tons chips (45.4 metric tons)
50% moisture

1200 ft^3 white liquor (34.0 m^3)
5.7 lb/ft3 effective alkali concentration (91.4 g/l)

1300 ft^3 black liquor (36.8 m^3)
Effective alkali concentration of black liquor can be neglected.

What is the effective alkali charge?

$$50 \text{ tons} \times 2000 \frac{\text{lbs}}{\text{ton}} \times 0.5 = 50{,}000 \text{ lbs. oven-dry wood}$$

$$1200 \text{ cu. ft.} \times 5.7 \frac{\text{lb}}{\text{cu. ft.}} = 6840 \text{ lb. effective alkali}$$

$$\text{Effective alkali charge} = \frac{6840}{50{,}000} \cdot 100 = 13.7\%$$

The cooking cycle

Presteaming

Before the chips are impregnated with cooking liquor, they are usually presteamed. The reasons for presteaming are: (1) to drive off the air inside the chips and replace it with steam or water; and (2) to heat the chips.

When presteaming is used in batch digesters, it normally takes place during chip filling. In continuous digesting there is a special presteaming vessel the chips pass through before entering the digester. If there are air pockets in the chips, the liquor will not penetrate as easily and the result may be uneven cooking.

Chip impregnation

After presteaming, cooking liquor is added to the chips and *impregnation* begins. The objective of impregnation is to distribute the cooking liquor uniformly into the chips. The impregnation consists of two different processes: bulk penetration and diffusion.

Bulk penetration is the movement of the cooking liquor through the pores of the wood.

Diffusion means that ions and molecules move from an area of high concentration to an area of lower concentration in order to even out differences. Diffusion of cooking chemicals can, therefore, only take place if the wood pores are already filled with liquor.

In batch digesting, impregnation takes place during the linear temperature rise to the cooking temperature. It is important that the impregnation is completed before the temperature rises above 130 °C since undesirable reactions can start in the absence of alkali. In addition, too rapid a heating rate will lead to non-uniform liquor and temperature distribution, resulting in an uneven cook with high reject levels.

In continuous cooking the impregnation takes place at a constant temperature below the level at which significant delignification occurs, usually at about 120 °C. For both batch and continuous cooking a major part of the alkali is actually consumed during the impregnation since most of the total loss of carbohydrates takes place here. In addition, most of the extractive fraction is removed during the early stage of impregnation.

Cooking

After impregnation is complete, it is time for the cooking stage. In both continuous and batch digesting, it takes place at a constant temperature of approximately 170 to 180 °C for one to two hours. Most of the delignification takes place in the cooking stage.

In continuous digesting, the cooking reactions are stopped by displacing the hot cooking liquor with cool washing liquor. A washing stage generally follows before blowing. In batch digesting the cooking is normally disrupted by lowering the pressure and temperature and subsequent blowing.

Blowing

During the whole cooking cycle the chips retain their structure, and it is not until the blowing process that the chips are fully separated into fibers, if the kappa number has been lowered sufficiently.

During blowing the temperature and pressure are lowered quickly. The result is that the liquor inside the chips starts to boil and the resulting pressure is enough to separate the fibers in the chip.

Variables affecting the cooking

The variables affecting the cooking process can be divided into three categories:

- Chip quality
- White liquor properties
- Cooking control variables.

Of these variables, cooking control variables are the only ones the operator has any control over when cooking. At that point, the chip quality and white liquor properties are already set.

The main cooking control variables are:

- Time and temperature (H-factor)
- Alkali charge
- Liquor-to-wood ratio.

Sulfidity is a variable the operator cannot control since it is a white liquor property. However, it can be changed by altering the ratio of sodium to sulfur in the makeup chemical, and is affected by washing efficiency, recovery boiler operations and by salt cake additions.

Time and temperature

The delignification reactions are very temperature dependent. A small increase in temperature has a big effect on the delignification rate. For example, an increase of 10°C from 160°C to 170°C will more than double the delignification rate.

Figure 4.4 shows the relative reaction rate as a function of the temperature. The relative reaction rate is arbitrarily set to 1.0 for 100°C. As one can see in Figure 4.4, the delignification reactions are very slow at low temperatures, but increase quickly at temperatures above 160°C.

Up to about 175°C the temperature does not affect anything else, except the delignification rate. When *higher* cooking temperatures are used, the process becomes less selective for lignin and more cellulose is degraded. This leads to a loss in yield. At temperatures above 190°C, the loss in yield and strength can be substantial due to excessive cellulose degradation. This is illustrated in Figure 4.5.

The cooking time is important, since the delignification reactions are so rapid at high temperatures. A few additional minutes during impregnation will probably not affect the final quality of the pulp, but a few additional minutes during cooking certainly will.

A method has been developed for treating cooking times and temperatures in kraft pulping as a single variable. Thus, the times and temperatures of any cooking cycle can be represented by one single numerical value, the so called *H-factor.*

Cooking cycles which give equal H-factors will produce pulps of equivalent yield and lignin content when other conditions are identical. The combination of time and temperature does not have any effect; as long as the H-factor is constant, the yield and lignin content will be the same.

The H-factor represents the area under a curve in which the relative reaction rate is plotted against time. When the time-temperature cycle is known, it is easy to plot relative reaction rate versus time. Figure 4.6a shows two time-temperature cycles. Figure 4.6b shows the corresponding relative reaction rate-time curves. Both time-temperature cycles result in an H-factor of 1300.

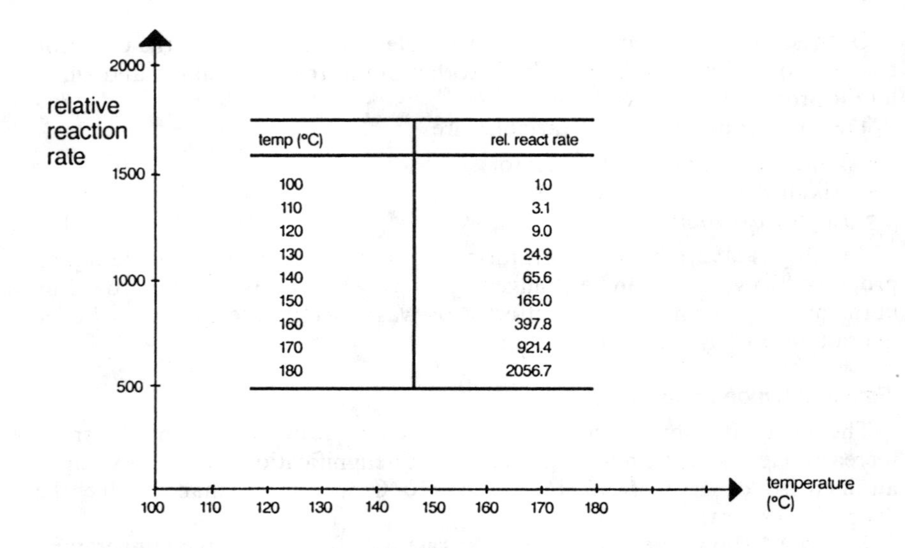

temp (°C)	rel. react rate
100	1.0
110	3.1
120	9.0
130	24.9
140	65.6
150	165.0
160	397.8
170	921.4
180	2056.7

Figure 4.4 Relative reaction rate vs. cooking temperature.

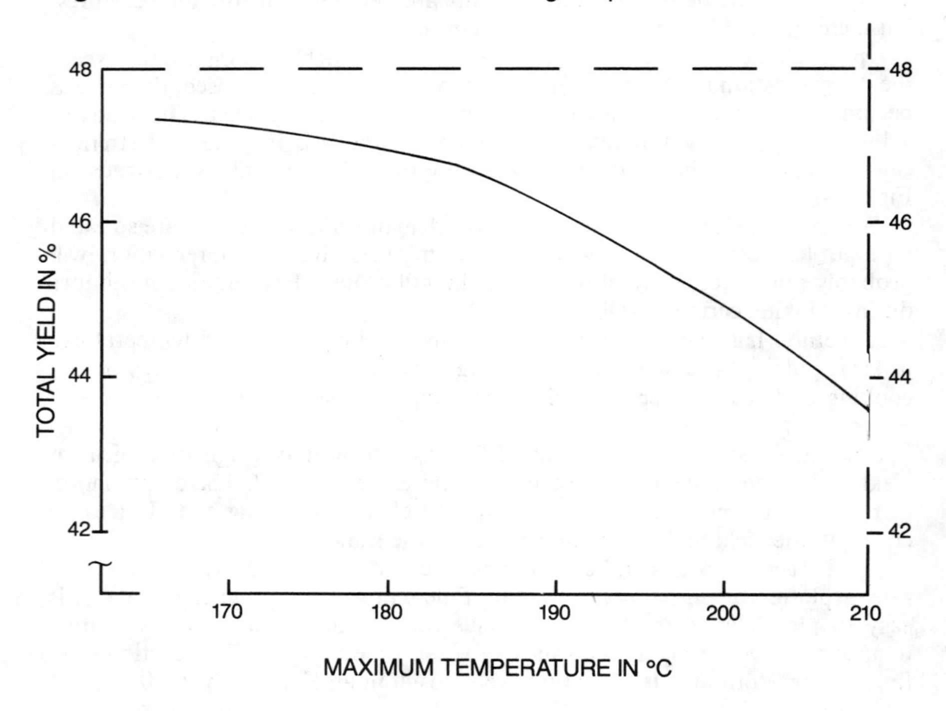

Figure 4.5 Total yield vs. maximum temperature.

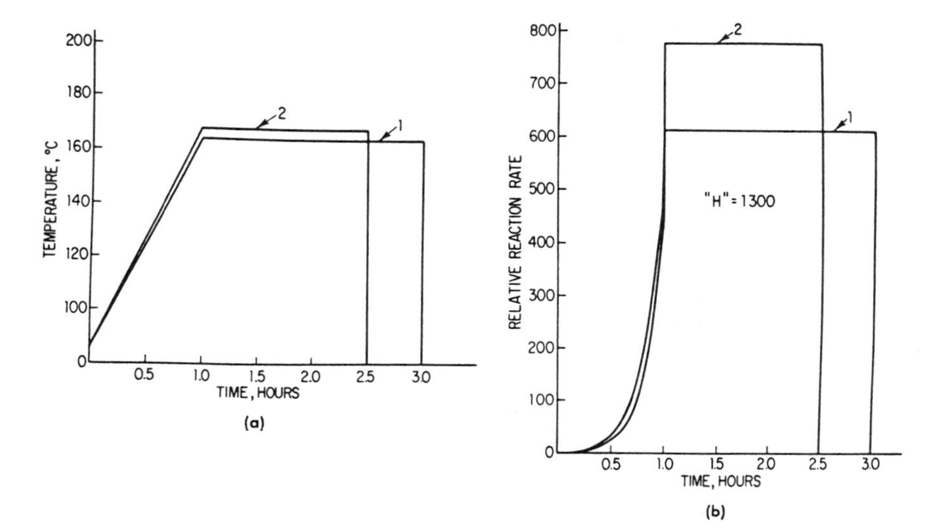

Figure 4.6 H-factor illustration.

As one can see, the contribution of the heating time to H-factor is very small compared to the contribution from cooking time at higher temperatures. Modern digester control systems automatically compute and accumulate H-factor during the cook to compensate for deviations from the intended cooking cycle.

Calculation of the H-factor

To calculate H-factor the area under the relative reaction rate-time curve must be numerically evaluated. The easiest way to do this is to divide the time-to-temperature part into time segments and use the average relative reaction rate in each segment.

The following is an example of how H-factor is calculated numerically:

Digester time-temperature cycle:
I: Heating from 90°C to 180°C during 1.5 hours
II: Cooking at constant temperature; 180°C, for 0.5 hours.

Relative reaction rate for temperatures 90–180°C

°C	relative reaction rate
90	0
105	2
120	9
135	41
150	165
165	608
180	2057

Average the relative reaction rate values over fifteen-minute periods in the heating part of the time-temperature cycle. The resulting calculations are:

time from start (hrs)	temp. (°C)	relative react. rate	average rel. react. rate	time period	H-factor contribution
0.0	90	0			
			1	× 0.25	0
0.25	105	2			
			6	× 0.25	1
0.50	120	9			
			25	× 0.25	6
0.75	135	41			
			103	× 0.25	26
1.00	150	165			
			386	× 0.25	97
1.25	165	608			
			1332	× 0.25	333
1.50	180	2057			
			2057	× 0.5	1028
2.00	180	2057			
				Total:	1491

The resulting H-factor is 1491. Depending on the mill, final H-factor values typically range from 1000 to 1500.

Alkali charge

The normal effective alkali charge ranges from 10% to 16% (as Na_2O on dry wood). It varies depending on wood species, cooking conditions, and degree of delignification required, as can be seen in this table:

Product	Kappa Number Range	Effective Alkali Charge (%)
bleachable hardwood	13 to 15	13.5 to 16.0
bleachable softwood	28 to 35	14.0 to 17.0
linerboard (softwood)	75 to 100	11.0 to 13.0

In order to complete the cook in a reasonable time, a small excess of cooking chemicals is usually charged, about 10% more than is consumed. This alkali excess also makes sure the pH does not sink below the level where dissolved lignin starts to redeposit onto the fibers.

A higher alkali charge will increase the rate of delignification. Therefore, by increasing the effective alkali (EA), one can cook to a lower H-factor and still reach the same kappa number. (See Figure 4.7.) An increase in EA charge, though, will also affect the yield. The amount of dissolved hemicellulose increases with increasing EA charge and the yield is reduced. This is illustrated in Figure 4.8.

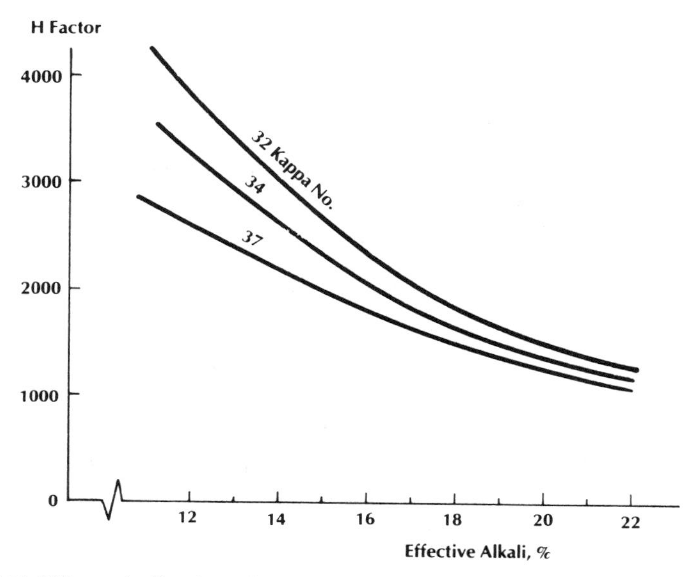

Figure 4.7 Effect of effective alkali charge and H-factor on kappa number.

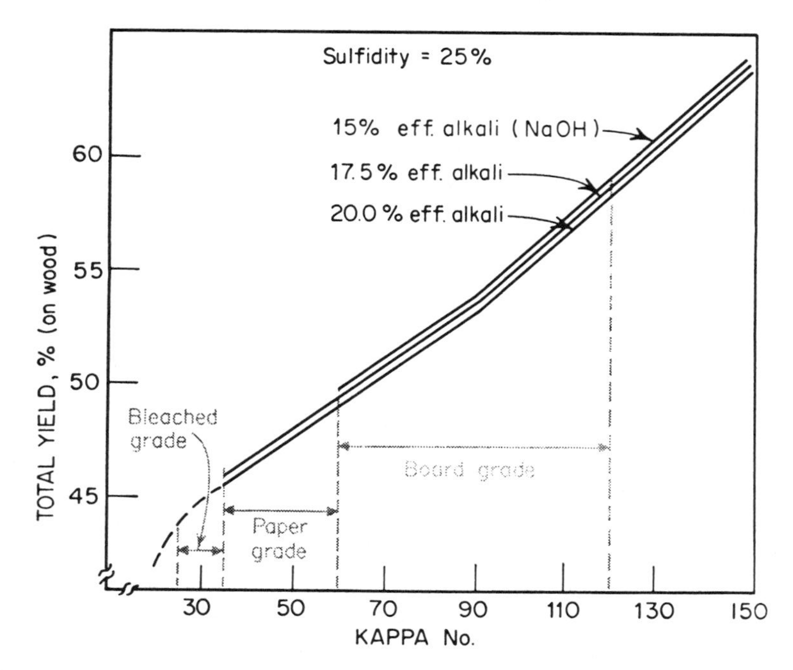

Figure 4.8 Pulp yield vs. kappa and effective alkali charge.

Liquor-to-wood ratio

In a batch digester, the white liquor required to meet the effective alkali charge is normally much less than the volume required to completely cover the chips. Black liquor is added to make up the difference. The more total liquor added to a given amount of chips, the higher the liquor-to-wood ratio. The normal range is from three to five.

In general, mills try to minimize the liquor-to-wood ratio by using some chip packing method. Well packed chips have less void space and need less added liquor to become fully immersed. Advantages of well packed chips and a low liquor-to-wood ratio are:

- Higher productivity due to increased wood charge per cook
- Lower costs for liquor heating
- Lower costs for liquor recovery since less water needs to be evaporated.

In continuous digesters liquor-to-wood ratio must be viewed from a slightly different standpoint. In normal operation no black liquor is added and the apparent liquor-to-wood ratio is quite low. The residence time of the cooking liquor in a continuous digester is shorter than that of the chips. Wash water added in the base of the continuous digester has much the same effect as black-liquor recycle in batch digesters. The effective liquor-to-wood ratio in the cooking zone is actually higher than that based on the white-liquor flow alone.

Sulfidity

Although not an operator-controlled variable, the effect of sulfidity on pulping is treated here. The presence of sodium sulfide in the cooking liquor results in a more rapid delignification, while the carbohydrate dissolution is more or less unaffected. This means that, compared to soda pulping, where only NaOH and no Na_2S is used, the pulping is faster and gives a stronger pulp at a higher yield since less carbohydrates are degraded and dissolved. The positive effects of sodium sulfide are quite dramatic up to 15% sulfidity but, then, the increases in pulp yield and strength start to level off.

The optimum sulfidity depends on several factors, such as wood species, alkali charge, cooking temperature, and the properties desired in the final product. The normal range is from 15% to 35%. Generally, the optimal sulfidity for hardwoods is lower than for softwoods, 20% compared to 25%. For production of coarser grades, such as linerboard which requires mostly softwood chips, a sulfidity of 20% is often found sufficient. When the sulfidity rises above 25%, the odor from the mill usually increases sharply. The odor release is often the factor that determines a mill's upper sulfidity level.

Sulfur is lost throughout the process: during digesting with the relief gases, during pulp washing, in the evaporators, and in the recovery boiler. It is mostly lost in the form of different gases such as SO_2, H_2S, methyl mercaptan, and other mercaptans. The release of hydrogen sulfide (H_2S) and the mercaptans

Control Parameters

The objective in kraft pulping is to dissolve as much lignin during cooking as required in order to reach a set maximum lignin content in the resulting pulp. One can also say that the objective is to cook to a *target kappa number.* The kappa number is a measure of how much lignin the pulp contains.

Until recently, no method existed that could measure the kappa number during cooking. Mills therefore had to be able to predict the kappa number based upon the cooking conditions.

The aforementioned H-factor is used in digester control. There are also modifications of the H-factor concept that can take variations in alkali charge and sulfidity into account as well.

The wood furnish will always affect the cooking and give variations in the cooking result even if H-factors and liquor additions are identical. This is because the wood furnish is never homogenous. It can vary in wood species, composition, bark content, moisture content, chip size distribution, and degree of decay.

The variations in wood quality and alkali charge will influence the alkali concentration during cooking. Some mills measure the *residual alkali*, which is the alkali concentration at the end of the cook. This reading of residual alkali makes it possible for the mill to compensate the H-factor for variations in wood quality and alkali charge.

The alkali concentration is measured in on-line alkali analyzers, generally located on a liquor circulation stream. There are a number of different measuring principles. Measurements of conductivity, refractive index, and color can all be related to alkali concentration.

Evaluation of pulp

Kappa number

The kappa number test determines the lignin content in pulp. It is used in mill control for two purposes:

1. To indicate the degree of delignification achieved during cooking; i.e., the kappa number is used for cooking control.
2. To indicate the chemical requirement for bleaching.

In the kappa number test, a known amount of permanganate solution is added to a pulp sample. After a certain time, the amount of permanganate that has reacted with the pulp is determined by titration of the sample. The kappa number then is defined as the number of ml of 0.1 N potassium permanganate solution consumed by 10 g of pulp in 10 minutes at 25°C. The results are corrected to 50% consumption of the permanganate added.

For kraft pulp the relationship between kappa number and lignin content is:

$$\text{lignin, \%} = 0.147 \times \text{kappa number}$$

The relationship between lignin content and kappa number is very good for bleachable pulps.

There are other, older methods for evaluating the lignin content in pulp; the permanganate number test is a predecessor to the kappa number test. The Roe number and the chlorine number are two tests that both use chlorine as a reagent when determining lignin content. They are not used as mill control tests any longer, since chlorine gas is hazardous and problematic to handle.

Viscosity

By dissolving a pulp sample in a cellulose solvent and then measuring the viscosity of the solution, one can get a good estimate of the degree of polymerization of the cellulose, i.e., the average cellulose chain length. A lower viscosity means a more degraded cellulose consisting of shorter cellulose chains. Once viscosity falls below a certain level, the pulp strength starts to decrease. Viscosity is often used as a measure of cellulose degradation during bleaching.

Drainability

The resistance of pulp fibers to the flow of water is an important consideration in pulp processing and papermaking. The most common method used for testing drainability is the Canadian Standard Freeness (CSF) tester. CSF is the amount of water (in mL) collected from the standard tester's side orifice when the pulp is drained through a perforated plate at 0.30% consistency at 20°C. The typical range for freeness is between 20 to 700 mL. Coarse pulps with high lignin contents normally drain faster than pulps with low lignin content. A fast drainage rate corresponds to a high CSF.

Beater evaluation

The most important property of pulp is its papermaking potential. To be able to evaluate this property, the pulp must first be treated mechanically by beating or refining. In the mill this is done in refiners, but in the lab a Valley beater or a PFI mill is used. The degree of beating is measured by doing Canadian Standard Freeness tests. A more beaten pulp drains more slowly than a less beaten pulp; i.e., the freeness decreases with the amount of beating.

After the pulp has been beaten, standardized handsheets are formed. These handsheets are then tested for different physical properties such as tear, burst, tensile, brightness, and opacity.

Questions

1. Define the objective of kraft pulping.
2. Name the active components in cooking liquor.
3. Discuss the reactions of lignin during kraft pulping.
4. Describe how cellulose and hemicellulose are degraded during pulping.
5. Understand the practice of expressing sodium compounds in cooking liquor as equivalent amounts of sodium oxide.
6. Give definitions for and calculate the following kraft pulping terms: total alkali, active alkali, effective alkali, and sulfidity.
7. Name some dead-load chemicals found in white liquor and identify their source.
8. Define and calculate "liquor-to-wood ratio."
9. Define and calculate "alkali charge."
10. List the different stages of the kraft pulping process.
11. Give one reason why presteaming chips before impregnation is desirable.
12. Name and describe the two different processes by which chips become impregnated with cooking liquor.
13. Specify a temperature range typical for the cooking stage of kraft pulping.
14. Describe what happens during the blowing process.
15. List the main cooking control variables.
16. Describe the effect of temperature on the delignification reactions.
17. Discuss the concept of H-factor.
18. Understand the basics for the calculation of H-factor.
19. Discuss the effect alkali charge has on the pulping reactions.
20. Explain how liquor-to-wood ratio can affect the profitability of a pulping operation.
21. Describe what effect sulfidity has on the pulping operation.
22. Identify two parameters often used for digester control.
23. Name some methods used for evaluation of pulp.
24. Explain what the term "kappa number" stands for.
25. Describe testing of viscosity and identify what property of pulp it measures.
26. List some pulp properties that can be evaluated by the making and testing of handsheets.

Principal Sources for Figures

Smook, G. A. *Handbook for Pulp and Paper Technologists,* Edited by M. J. Kocurek. Atlanta and Montreal: Joint Textbook Committee of the Paper Industry, 1982.

TAPPI, *Introduction to Pulping Technology,* TAPPI Home Study Course, no.2. Atlanta: TAPPI PRESS, 1976.

Fig. 4.1 TAPPI, Fig. 6.4, p. VI-5.
Fig. 4.4 TAPPI, Fig. 12.13, p. XII-9.
Fig. 4.5 Smook, Fig. 7.11, p. 73.
Fig. 4.6 Smook, Fig. 7.12, p. 73.

5

Kraft Pulping: Equipment

There are two basic digester designs: batch digesters and continuous digesters.

Batch digesters fall into two categories: directly heated and indirectly heated. The principal operations in batch digesting include chip packing and steaming, liquor filling, slow temperature rise to assure complete penetration of the chips by liquor, relief of gases, cooking at maximum temperatures, relief of pressure, and blowing the digester. Each of these operations affects pulp properties and variations in quality.

Continuous digesters, as represented by the Kamyr system, separate the principal operations between different vessels, and between different sections within the digester. Chip charging and liquor circulation patterns are different. In digesters, washing is also standard in these continuous units. Liquor is removed earlier from the Kamyr, while the chips are removed at the base along with wash liquor.

Batch Digesters

Construction

Shape and dimensions

In earlier days several different kinds of pressure vessels were used for batch cooking, including horizontal and spherical vessels and rotating vertical vessels. Currently, batch digesters are stationary, vertical cylinders with a cone-shaped bottom part to make them easier to blow.

Batch digesters range from 70 m^3 to 340 m^3 capacity, with a standard capacity of 170 m^3 to 230 m^3 for most modern mills. Large digesters are more efficient, because they require less labor and use less steam per unit of pulp than small ones. There is a practical upper limit to unit capacity, though. Larger units require thicker walls, which increases production costs. Production, transportation, and installation all become more difficult with increasing unit size. Also, a mill would not want its total capacity tied to two or three huge digesters because the shutdown of one digester would tie up too great a proportion of the mill production.

Modern digesters are equipped with automatic capping and blowing valves. However, there are still old digesters in use where the capping head has to be bolted on manually after the digester has been filled with chips and liquor.

Construction material

Generally, kraft digesters are manufactured of carbon steel because the alkaline kraft cooking liquors are much less corrosive than sulfite liquors. Sulfite digesters are lined with acid resistant brick to withstand the corrosion.

Today the standard method of construction for batch digesters is a shell, completely welded of 2-inch (51 mm) steel plate. Bottom sections are sometimes reinforced by using 3-inch (76 mm) plate.

The average life span for a carbon steel kraft digester is about 20 years. The carbon steel is not completely resistant to the kraft liquor, however, and an annual corrosion of the walls of 0.5 mm to 1.0 mm is normal.

There are two types of stainless steel protection for carbon steel digesters: stainless steel lining installed in existing digesters and stainless steel cladding bonded to carbon steel plate in the rolling mill where the steel plates are manufactured. The corrosion rates for stainless steel clad digesters are about 10% of the corrosion rates for carbon steel.

Heating systems

A batch digester can be heated in two different ways, by direct steam or by indirect heating with forced circulation.

Direct steam is the simplest way of heating. With this method, steam is injected through a valve in the bottom of the digester. The difference in temperature between the top and bottom makes the liquor circulate by convection, and hot liquor rises through the middle of the digester, while colder liquor at the top flows down the walls to the bottom where it meets hot steam and is reheated. (See Figure 5.1.) An external liquor circulation system is sometimes present to increase uniformity.

The equipment for direct steam heating is little more than a steam line and a valve. Indirect heating requires more equipment, including a circulation system with a pump, an external heat exchanger, and a strainer section in the digester walls. (See Figure 5.2.)

In a common design, the liquor is removed from the digester through strainer plates located in the middle section. The liquor is heated in a heat exchanger where the heating medium is steam that condenses. The hot liquor is then returned to the top and bottom of the digester. The circulating pump is usually sized to turn over the liquor once every 10 minutes.

Direct heating has the advantage of being a very simple means to provide rapid heating. There are disadvantages, however. The cooking liquor becomes diluted with steam condensate, putting an additional load on the evaporators and resulting in a low economy of recovery. Also, the heating is non-uniform, resulting in temperature differences up to 10 °C in large digesters. Non-uniform heating results in uneven cooking that lowers the quality of the pulp.

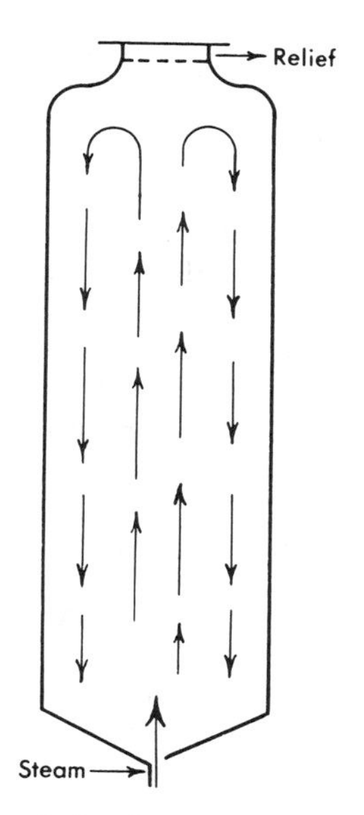

Figure 5.1 Convection circulation in a digester.

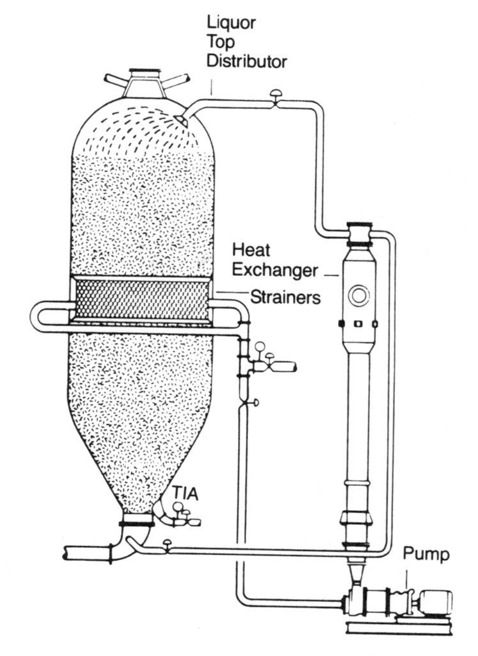

Figure 5.2 Indirect heating system in a batch digester.

Indirect heating with forced liquor circulation avoids liquor dilution and a more uniform temperature profile throughout the digester is achieved.

Chip and liquor filling

There are several routines for loading a batch digester with chips and liquor. The most common method is to add the chips and liquor simultaneously. The liquor lubricates the chip surfaces and allows them to pack well. This method allows fairly good packing density while keeping the time required for digester filling to a minimum. This method does have some disadvantages, however.

A higher chip packing density can be achieved by using steam packers or mechanical packers. As high a chip packing density as possible is desired since the wood charge per cook is increased, raising total production. Also, by simultaneous filling of chips and liquor, presteaming of the chips is not possible unless it is done outside the digester.

Both steam packers and mechanical packers give the chips a tangential motion when they enter the digester. The chips then drop in a spiral motion, spread evenly, and form a flat bed rather than a bed with a conical profile as is the case with no chip packing. This is illustrated in Figure 5.3. Uniform packing is important since uneven packing will lead to uneven liquor circulation, and allow the liquor to find "shortcuts" through the chips mass.

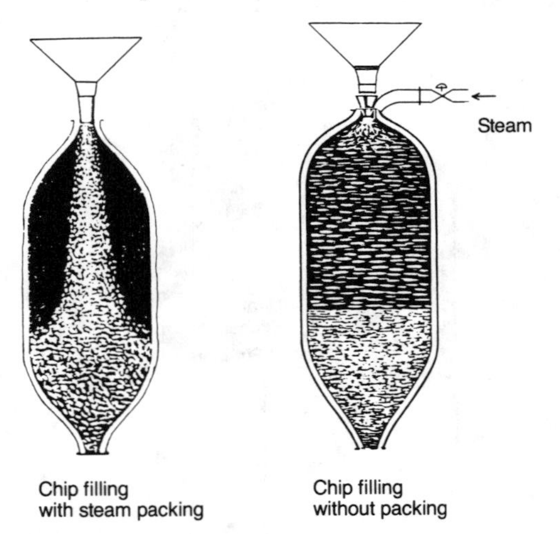

Figure 5.3 Chip packing system.

The digester charge can be increased up to 40% by using steam or mechanical packing. Special arrangements for blowing the pulp may be needed for extremely high chip packing density, however.

If presteaming is used, it is generally done during chip filling. If presteaming is not complete after filling, additional steaming can take place by injecting steam in the bottom of the digester. The purpose of presteaming is to drive off the air trapped in the chips and replace it with steam. Air remaining in the chips will hinder proper liquor impregnation and result in uneven cooking and pulp quality. While steaming, the purged air must be removed from the digester. Saturated steam of atmospheric pressure is usually used for steam packing and presteaming.

Heating and cooking

The choice of heating rate, cooking time, and cooking temperature varies from mill to mill even if they produce the same pulp product. This is because each mill has its own constraints limiting the possible ranges for these variables. For example, maximum allowable pressure for the digesters can limit the cooking temperature and a small heat-exchanger area in the liquor circulation system can be the factor that determines the heating rate.

There are some general rules, though. The heating time to reach maximum temperature is generally considerably longer than the time at maximum temperature. This is because it is important that good penetration and uniform heat distribution throughout the digester be achieved before reaching the cooking temperature. Another reason is limited steam capacity of the power plant, which puts a limit to the heating rate.

Normally, the cooking liquor is introduced to the digester at a temperature of about 70°C. Heating time varies widely from 30 to 120 minutes. Maximum temperature reached (cooking temperature) ranges from 160°C to 180°C, and the time at temperature (cooking time) can vary from 30 to 90 minutes. The digester spacing, i.e., the time it takes to blow the digester then fill it up with new chips and liquor, is normally from 20 to 40 minutes.

Altogether, the cover-to-cover time for kraft cooking cycles ranges from 2.5 to 4 hours, but can be less than two hours for packaging grade pulps. Figure 5.4 shows a typical cooking schedule for a batch digester. Typically, the time cycle for each cook is kept the same as much as possible in order to maintain the proper staggered sequence between the different cooks.

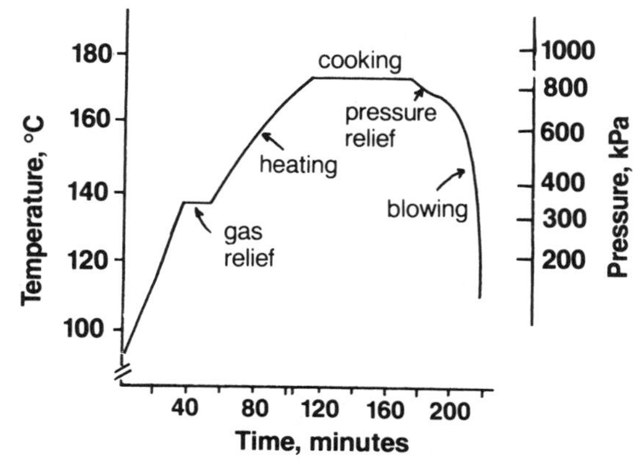

Figure 5.4 Cooking schedule for batch digester.

Digester relief

During heating and cooking, gases are formed from the extractives in the wood. Remaining air and other noncondensible gases, such as the carbon dioxide that is released in the cooking reactions, also accumulate. These gases must be removed from the digester. Otherwise the digester pressure will be higher than the steam pressure corresponding to the liquor temperature. This "false" pressure can lead to problems with cooking control.

Normally gas relief takes place during heating. Once a predetermined pressure is reached, a relief valve is opened. The accumulated gases are released, the valve is closed, and the heating continued. The procedure is repeated, if needed. The gas relief can be controlled, either manually or automatically. It is important that as little cooking liquor as possible follows the gas out, since this would lower the alkali charge in the digester. The relief gases go to a turpentine recovery system where raw turpentine is recovered.

Blowing

When the predetermined cooking time is reached, some of the pressure is released by opening the gas relief valve and releasing gases. These relief gases contain much less turpentine than the previous relief gases and can be treated separately to avoid mixing; but normally they go to the turpentine recovery system as well. Once the digester has reached blowing pressure, the bottom valve is opened and the content is blown into a blow tank. The blow tank is normally constructed of steel and is large enough to hold at least one and one half cooks.

The rapid temperature drop during blowing causes the liquor inside the chips to start boiling and evaporate to steam. The steam forces the chips to "explode" and the fibers are effectively separated. Pulp strength is adversely affected by harsh blowing of lower yield bleachable grade pulps.

The pulp slurry enters the top of the blow tank tangentially. A cyclone action ensures the blow steam is well separated from the stock. Blow tanks usually have baffle separators and large vapor spaces to achieve good separation and minimize carryover of liquor with the steam.

A considerable amount of steam is flashed off when blowing. When a digester is blown from 825 kPa (~120 psig), about one ton of steam per ton of pulp is produced. This heat must be recovered in order to achieve a good energy economy. In addition, the flash steam contains foul smelling gases that need to be destroyed.

A typical modern kraft batch digester blow heat recovery system is shown in Figure 5.5. The following is a description of how this blow heat recovery system works.

Flash steam and cool condensate pumped from the bottom of the accumulator tank enter the direct contact condenser (2). The steam condenses, producing hot "contaminated" water that flows to the top of the accumulator tank (4). From the top of the accumulator tank, hot condensate is pumped

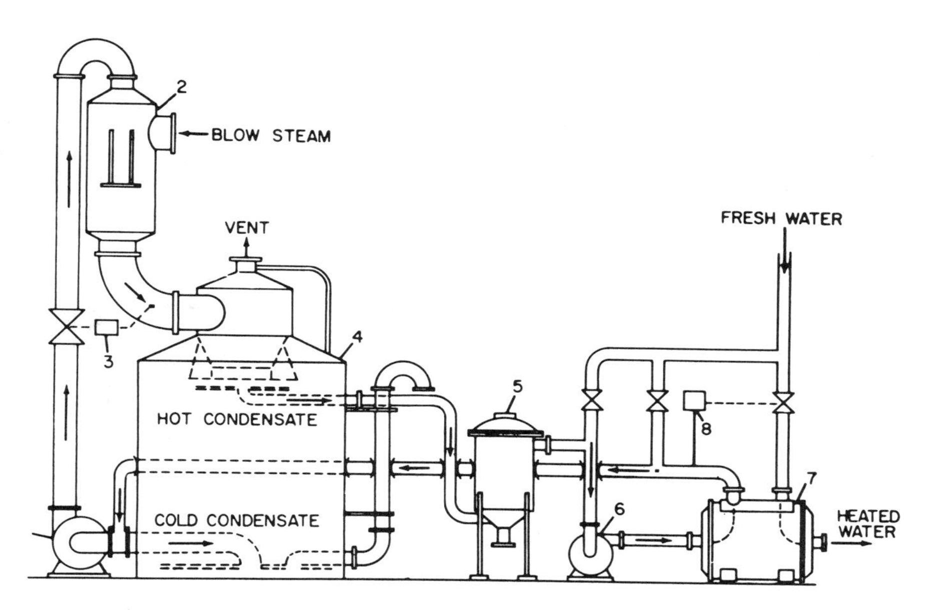

Figure 5.5 Blow heat recovery system for batch digester.

through a heat exchanger (7) that heats fresh water for pulp washing. The cooled condensate returns to the bottom of the accumulator.

The principle for the accumulator is that hot water has a lower density than cold water and, therefore, remains on top.

During a blow, a large volume of hot contaminated water accumulates in the top portion of the accumulator. Between blows, the hot condensate is "used up" in the heat exchanger and the interface between hot and cold layers rises. This moving interface between hot and cold sections provides for the accumulation capacity. Heat from each blow can this way be stored and then used continuously for heating fresh water. Some of the cold condensate is continuously removed from the accumulator to provide room for the new, hot condensate formed during each blow. The cold condensate is very dirty and is usually stripped with steam to remove methanol, sulfur-containing compounds, and other volatiles. The distillate is burned in the lime kiln to destroy the reduced sulfur compounds and recover the heat value from the organic components.

Cold blowing

The newest way to recover heat from batch digesting is cold blow cooking. This method is capable of lowering the steam demand of a batch cook by 40% to 50%, down to a level equal to that of continuous digesting. Also, pollution control is made easier by virtually eliminating flashing between cooks. A cold blow system is shown in Figure 5.6. This is how it works:

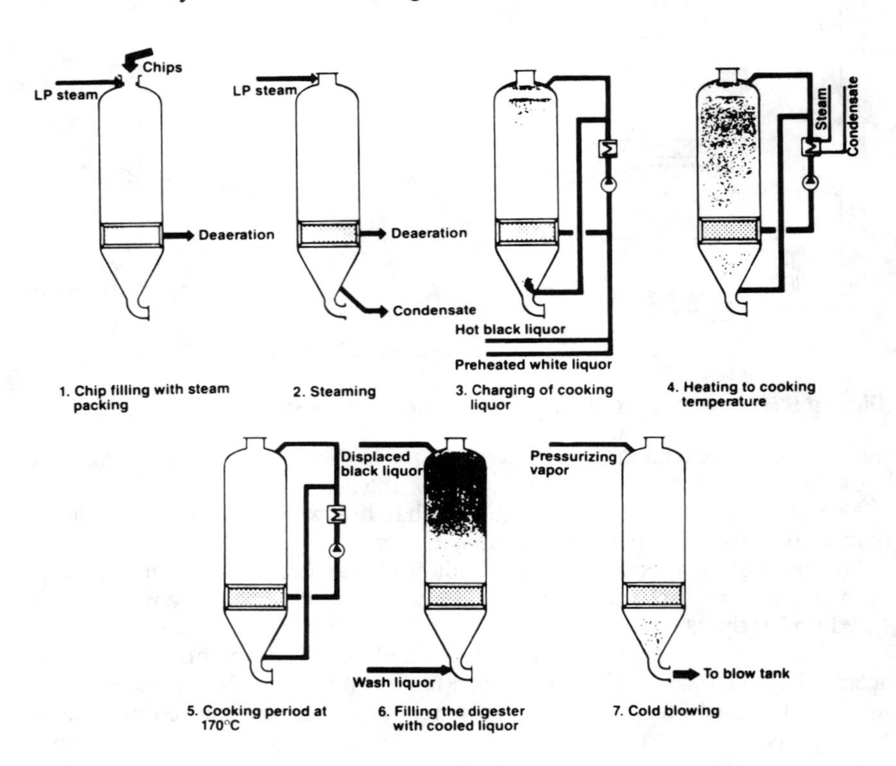

Figure 5.6 Cold blow system.

When the cook is complete, washing filtrate from washers is pumped into the bottom of the digester. It displaces the hot black liquor that is pumped from the top into a hot black liquor accumulator. About 75% of the black liquor is displaced before blowing. The blow valve is then opened and the digester is emptied. Because the temperature has been decreased before blowing, the internal pressure is low and some kind of external pressure has to be applied to the digester in order to remove the pulp. Hot black liquor from the accumulator tank preheats white liquor before it is pumped to the evaporators.

After blowing, the digester is filled with chips packed with a steam packer and presteamed. Then hot black liquor from the accumulator and the preheated white liquor is charged to the digester at a temperature of 135°C to 145°C. With this high starting temperature, the rise to cooking temperature can take place at up to 50% less time than usual. This is possible because the chips are presteamed, improving impregnation. Here is also where energy saving is achieved; much less steam is required for heating the cooking liquor.

By transferring most of the black liquor to the accumulator, the odorous gases remain in the liquor and are fed to the evaporators instead of being flashed off with the steam, which is what happens during hot blowing. Some steam and gases are released during cold blowing as well, but they require a much smaller and simpler condenser system.

Finally, it has been shown that pulp viscosity and strength are improved when using a cold blow technique. The forces acting on the pulp fibers are less violent than when blowing from high pressures.

Continuous Digesters

Introduction

By far the most widely used continuous digester design is the Kamyr digester. To date 330 units have been installed. Another similar design is the IMPCO (previously Esco) digester. Some 15 units are in operation.

The M&D digester differs from the Kamyr and IMPCO designs in that it is not standing upright but mounted at an angle of 45°. The charge is moved through the digester by means of an internal chain-driven conveyor. The M&D digester is mostly used for sawdust pulping.

Construction

The Kamyr continuous digester is a tall cylinder with a bottom section wider than its top section. The digester is a pressure vessel constructed of carbon steel. A normal height is 60 m to 70 m and an average capacity is about 1000 oven-dry tons (ODT) pulp per day.

Liquor recirculation streams play a major role in controlling the cooking process in the digester. The liquor is collected through screens on the digester periphery, circulated through a heat exchanger, then returned to the digester through a center pipe. There are different screens and pipes for each circulation flow.

Besides the digester, the system includes other equipment: a chip meter, a presteaming vessel, a chip chute, chip feeders, liquor heaters, a flash tank, and a blow unit. Figure 5.7 shows a typical Kamyr digester system.

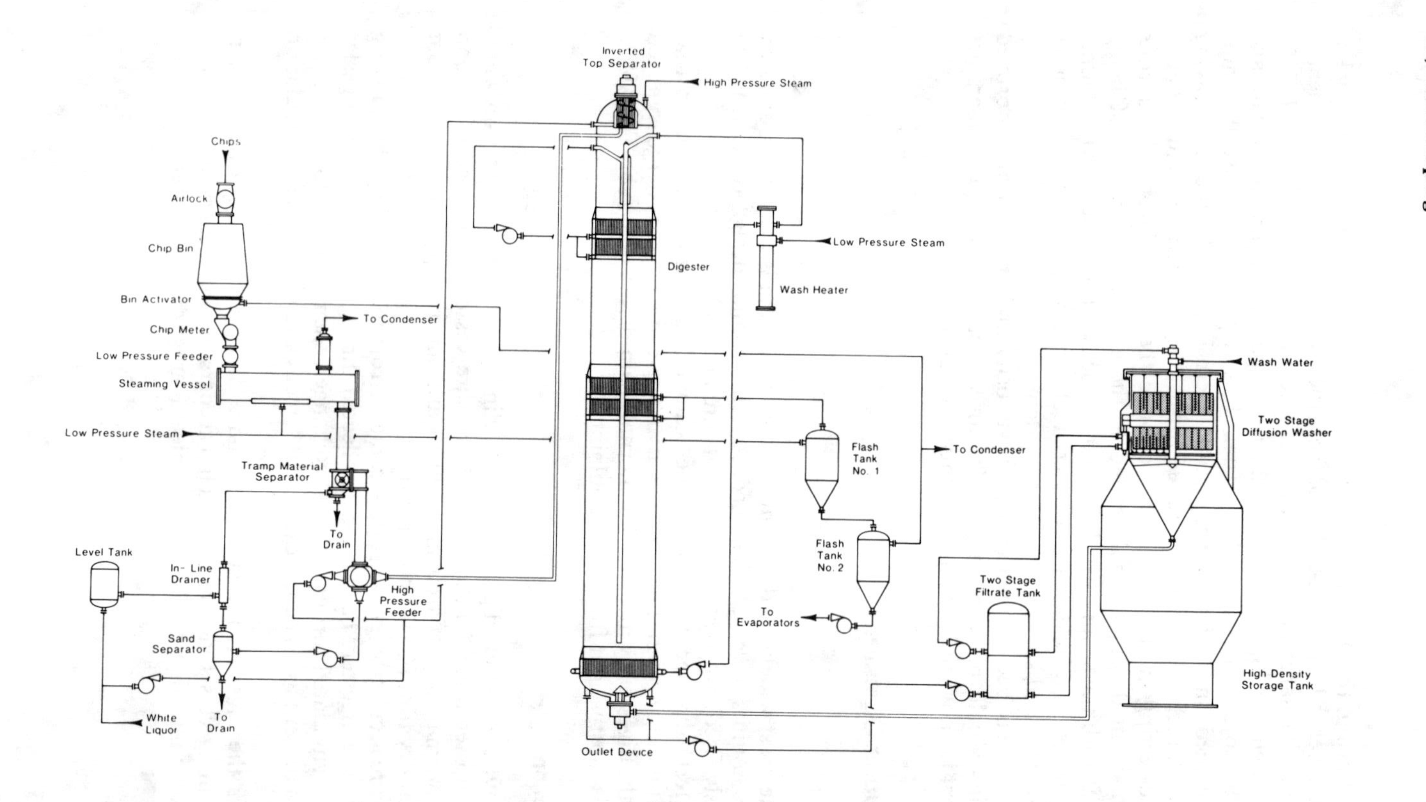

Figure 5.7 Single vessel steam-liquid phase digester with two-stage diffuser (Kamyr).

Chips and liquor charging

It is important that contaminants like tramp metal and rocks be removed from the chips before they enter the chip feeding system since it can be easily damaged.

Chip flow to the digester is controlled by the chip meter. It consists of a rotating wheel with pockets that fill up with chips. The diameter of the chip meter and the meter speed determine the volumetric flow of chips to the digester and the pulp production. All other flows must be adjusted to the rpm setting of the chip meter.

From the chip meter the chips pass through the rotating low pressure feeder into the *presteaming vessel.* Steam is introduced into one end of the vessel and noncondensable gases, air, and turpentine are vented from the other end. The required steam comes from the flash tank where spent cooking liquor from the digester is flashed. The chips stay in the presteaming vessel for about two to five minutes. The steam pressure is kept at around 200 kPa and the temperature is about 120°C. The purpose of this presteaming is to remove air from the chips and to heat them.

After presteaming, the chips pass through a chip chute where they are mixed with white liquor before passing to the high pressure feeder. The high pressure feeder is a rotary feeder totally submerged in liquor. It has pockets that become filled with chips and liquor from the chip chute. As the feeder rotates, the chips are sluiced away with the liquor in the high pressure top circulation flow, which has the same temperature and pressure as the top of the digester. This flow enters the top of the digester, which is completely filled with liquor. Here, the chips and liquor are separated by a screw surrounded by a strainer. (See Figure 5.8.)

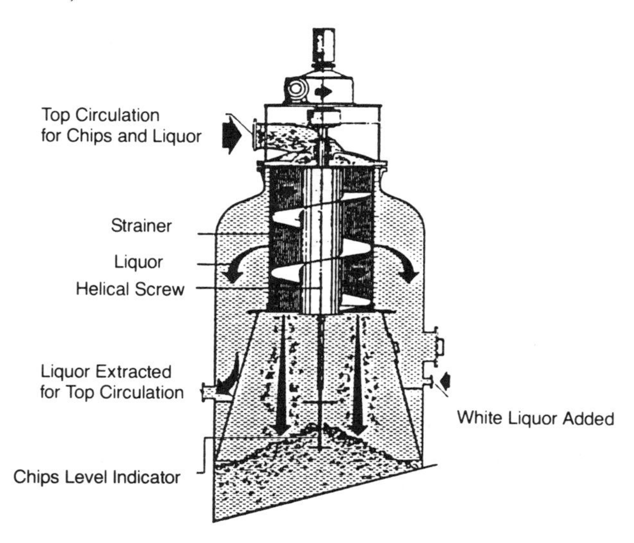

Figure 5.8 Top of continouous digester system.

The chips settle to form a plug that moves slowly down through the digester. The liquor required for the top circulation is extracted from the strainers at the top of the digester. Fresh white liquor is also added to the top of the digester.

Under normal conditions, there is no need to add black liquor to dilute the white liquor because the retention time for the liquor in the digester is about double the retention time of the chips. The alkali charge, therefore, works out to be the same as for batch digesting, where about 50% of the cooking liquor consists of black liquor. If the chip moisture is very low, however, black liquor can be added to ensure the digester is completely filled.

The impregnation and cooking zones

From top to bottom, the digester has four different zones: an impregnation zone, a small heating zone, a cooking zone, and a washing zone. (See Figure 5.9.) More modern designs have a separate vessel for impregnation. Most digesters have a thicker bottom section in order to allow for a longer washing time. With a larger cross section area, it takes longer for a certain amount of chips to flow through.

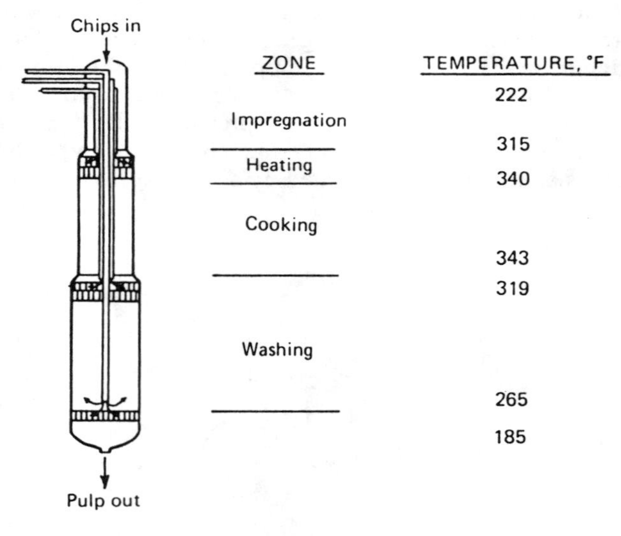

Figure 5.9 Zones in a continuous digester.

The chips and the liquor first descend through the impregnation zone. The impregnation takes place at 105°C to 130°C for about 45 minutes. The next zone is the heating zone, where the temperature is rapidly increased to the final cooking temperature. This is done in two steps by two liquor circulation systems. In the upper heating circulation, liquor is removed from the digester through a strainer in the digester wall and the liquor is pumped through an external heat exchanger where it is heated with steam. It is then returned to the heating zone through a center pipe in the digester. The liquor in the lower heating circulation is withdrawn through a strainer located just below the upper strainer. It is pumped through a second heat exchanger and is returned to the digester at a temperature of about 160°C to 170°C.

After this rapid heating, the chips pass through the cooking zone. This takes one to two hours. Since the pulping reactions are exothermic, a temperature rise of about 4°C above that provided by external heating takes place in the cooking zone.

At the end of the cooking zone the temperature must be lowered rapidly to stop the pulping reactions. This is done by the quenching circulation strainer. Washing liquor is drawn from strainers located at the top of the washing zone and is then pumped through a center pipe that ends at the bottom of the cooking zone. The washing liquor has a temperature of about 130°C. It can cool the chip mass rapidly. The spent cooking liquor is removed from the digester through a strainer located between the cooking and washing zones.

The cooking liquor is reduced to atmospheric pressure in one or two flash tanks. In a two-stage flash system, the first flash tank generates steam of about 120°C for presteaming of the chips. Steam from the second flash tank is used for either atmospheric presteaming of the chips in the chip bin or for hot water production. After flashing, the black liquor is pumped to the evaporators and further recovery.

The washing zone

The bottom part of the Kamyr digester contains the washing zone. Wash liquor, either hot water or filtrate from external washers, is pumped into the bottom of the digester at a temperature of about 80°C. Near the bottom there is an external recirculation loop where the wash liquor is heated to 130°C to 135°C. The wash liquor moves upward, countercurrent to the chip flow. The wash liquor is removed through the quenching circulation strainer, circulated, and injected to the pulp mass to lower the temperature. (See Figure 5.10.)

Finally, the wash liquor is removed with the spent liquor traveling downward with the chips for flashing and recovery. Consequently, the countercurrent flow in the washing zone is identical to the dilution of the system.

The retention time in the washing zone depends on the design of the digester and varies from one to four hours. Basically, washing in a continuous digester is a diffusion process where dissolved lignin and cooking chemicals spread from the chips into the surrounding liquor. Factors affecting the diffusion process are time, temperature, chip thickness, and amount of wash liquor used. A long retention time and high temperature improve diffusion. Chip thickness is an important variable since doubling the thickness will reduce the rate of diffusion by a factor of four.

A typical countercurrent digester wash zone with 1.5 hours residence time should be approximately equivalent to two stages of rotary filter washing at the same dilution factor.

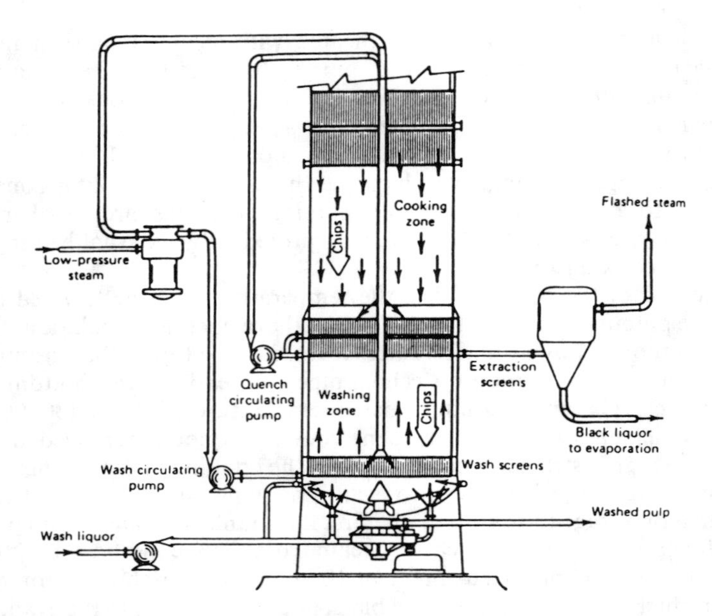

Figure 5.10 Washing zone with circulation systems.

A continuous digester operates under a hydrostatic pressure well above the steam pressure corresponding to the cooking temperature. This is to make sure that no flashing of liquor occurs at pump inlets and that the digester is totally filled with liquor at all times. The high pressure also assures good impregnation in the impregnation zone. A normal pressure is about 1240 kPa in the heating zone and around 1480 kPa at the bottom.

Blowing

When the chips finally reach the bottom of the digester they have a temperature of about 80°C. There they are continuously discharged by means of a bottom scraper. The bottom scraper consists of two radial rotating arms with scrapers mounted on them.

The blowing is usually done in steps since the pressure is so high. The blowing arrangement usually incorporates a control valve on either side of a small pressure vessel with a variable orifice so that there are three stages of pressure drop. For continuous digesters the blow tank is normally designed to hold pulp for 30 minutes to one hour.

Modifications to the Kamyr digester

Theoretical studies at the Swedish Forest Products Laboratory demonstrated that improved kraft pulping could be obtained by introduction of several modifications to the process. These were: a) low initial alkali concentration;

b) high sulfidity at the outset of the bulk phase of delignification; and c) reduced concentration of dissolved lignin and sodium salts in the later part of the cook. Kamyr developed modifications to the continuous digester to produce pulp under the conditions specified in the STFI studies. The modified process is shown in Fig. 5.13. The white liquor addition is split among the feed, the transfer circulation and a new countercurrent cooking zone. This split addition provides a more uniform alkali concentration over the entire cook. The countercurrent zone produces a low concentration of lignin and sodium salts in the later part of the cook. These modifications provide a higher viscosity, easier bleaching pulp in comparison to the pulp produced by the conventional continuous process. The modified process can be used to cook to lower kappa without loss in strength. Most Kamyr systems since 1986 have this feature incorporated in the design, or can be converted to include it.

The original digester had been designed exclusively for a practical kraft cook of the classical type with very little time allowance for presteaming and impregnation of chips and maximum production rate. Introducing a chip-liquor slurry into the high-pressure environment of the digester was to accomplish impregnation, which it did to a degree. The extensive circulation system of strainers and concentric ducts was sensitive to chip fragments and fines. It also caused mixing of hot used liquor with fresh cool liquor in the impregnation zone.

Although the hydraulic model is still the most widely used, the steam-liquid phase model (Fig. 5.11), which was developed primarily for cooking acid bisulfite and bisulfite pulps, has become the most versatile continuous digester, successfully in use also to cook prehydrolyzed kraft [6], NSSC, and regular kraft pulps.

The main modification was to replace the top separator of the hydraulic system by an inverted top separator and thereby maintain a vapor space in the top of the digester in which the impregnated chips can be heated rapidly to cooking temperature with direct steam. This means that there is no special impregnation zone in the digester, but in cases where good control over pulp quality is necessary, a small circulation loop is added in the upper section for temperature control. This circulation can also be used to introduce chemicals for maintaining a second cooking stage.

In the inverted top separator (Figure 5.12), top circulation liquor is drained off through vertical slots of the cylindrical screen while in the screw conveyor moves the chips and any surplus liquor up and over the rim of the separator where they are blown by direct steam into the digester. This provides uniform heating of the charge by the steam flow which is controlled by the temperature of the vapor phase.

The next development in the Kamyr continuous digester design was the improvement of presteaming and impregnation of chips in a separate vessel. These modifications provided an increased flexibility in process conditions and improved control over the impregnation and cooking stages.

In the two-vessel system (Figure 5.13), the presteamed chips are added through the high-pressure feeder into the top of the downflow impregnation

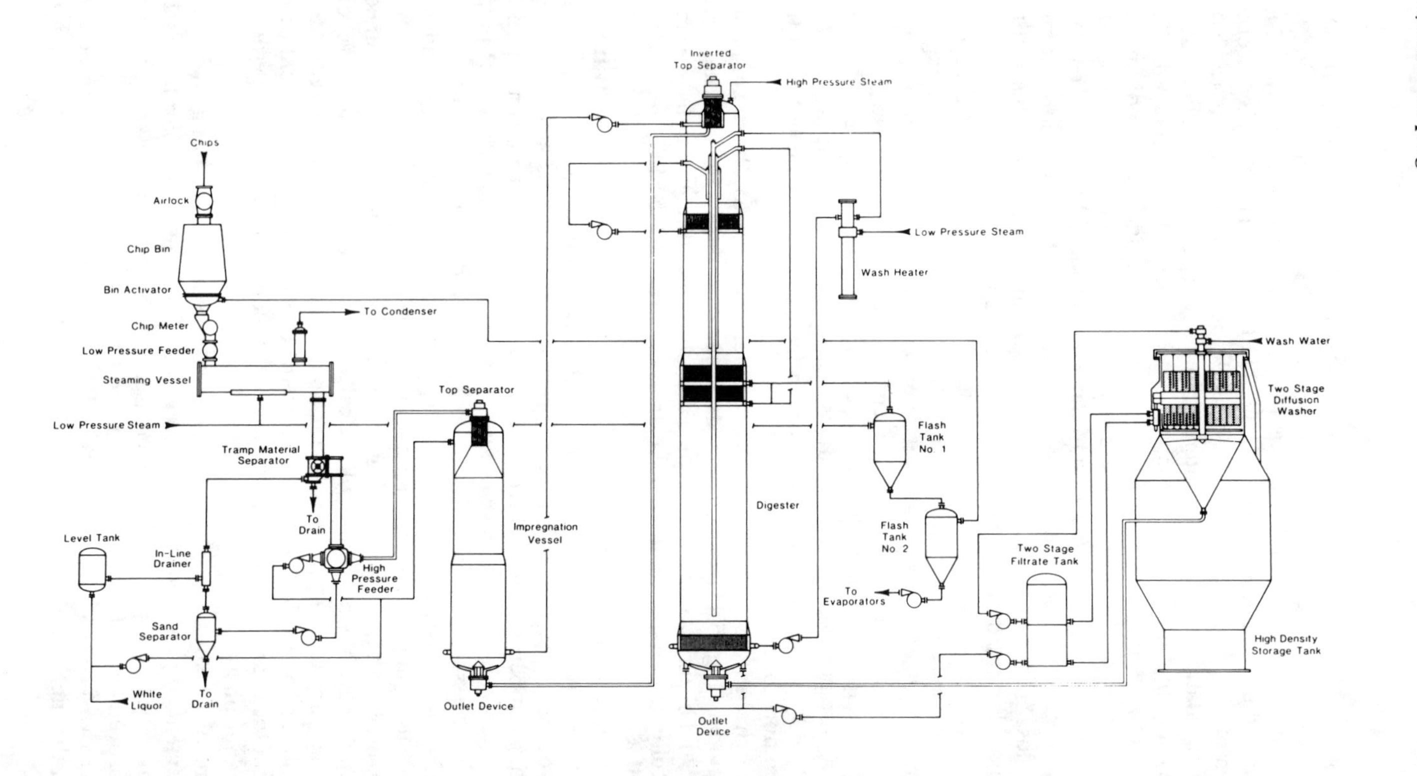

Figure 5.11 Two-vessel vapor liquor phase with two-stage diffuser (Kamyr).

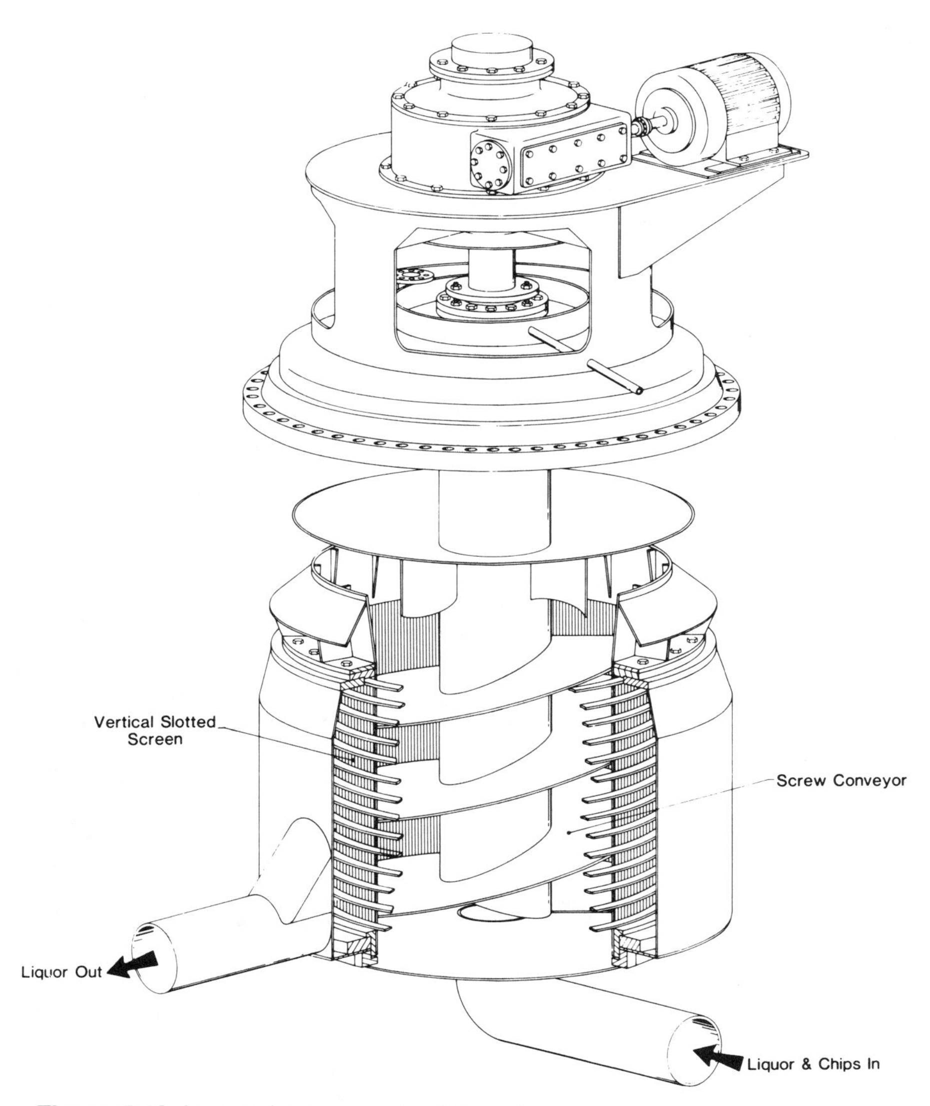

Figure 5.12 Inverted top separator (Kamyr).

vessel which operates at digester pressure plus hydrostatic head of the liquid column at 1400 kPa and 110° C. These conditions of high pressure and relatively low temperature provide good impregnation of various chip size fractions required for pulps with low rejects and high uniformity. From the outlet device of the impregnator chips and liquor are transferred by the sluice pump to the top of the digester with inverted top separator as described earlier.

Other continuous digesters

The IMPCO digester is a vertical down-flow, continuous digester fairly similar to the Kamyr digester in design and operation, but with some distinctive differences. An IMPCO digester can be seen in Figure 5.13.

The components of the IMPCO digester are all arranged vertically, with the chips feeding and pretreatment mechanism on top of the digester vessel. This arrangement makes placement of equipment less flexible and limits size.

The liquor circulation system has a somewhat different design than in the Kamyr digester. It has self-cleaning liquor strainers and multiple circulation liquor pipes. Also, the liquor and the chips flow in the same direction throughout the digester, which results in a uniform mass flow without risk of "hang-ups." No diffusion washing stage is thus included in the IMPCO digester.

Compared to the Kamyr digester, the IMPCO design is more tolerant of fines and sawdust in the furnish due to the improved liquor extraction and circulation.

The Messing and Durkee (M&D) digester consists of a large tube installed at an angle of 45°. (See Figure 5.14.) The tube is divided into an upper and lower compartment by a mid-feather, and is equipped with a chain conveyor that moves the chips by means of flights. The chips enter the system close to the top of the tube. They are then moved by the flights down the upper half of the tube and then up the lower half of the tube before being discharged.

The M&D tube can be divided into four treatment areas as far as liquor temperature or chemical concentration are concerned. The retention time can be precisely controlled by the speed of the chain conveyor. Thus, it is possible to maintain good control over the cooking conditions.

There are several industrial installations consisting of two M&D tubes in a series connected by a rotary pressure valve. This arrangement permits cooking with a great variety of treatment stages and better cooking control. Figure 5.15 shows a double-tube arrangement where the cooking stage is carried out in a steam atmosphere in the second tube.

The M&D digester is the only continuous digester where the cooking material is moved by mechanical means through the cooking liquor. Because of the mechanical movement of material, the digester can handle all ranges of chips quality and also sawdust. As a matter of fact, most commercially-installed M&D systems are used for pulping sawdust.

One disadvantage of the M&D digester is its small size. Since it is shop-fabricated, the maximum size of an individual vessel is limited to a diameter of about 2.5 m.

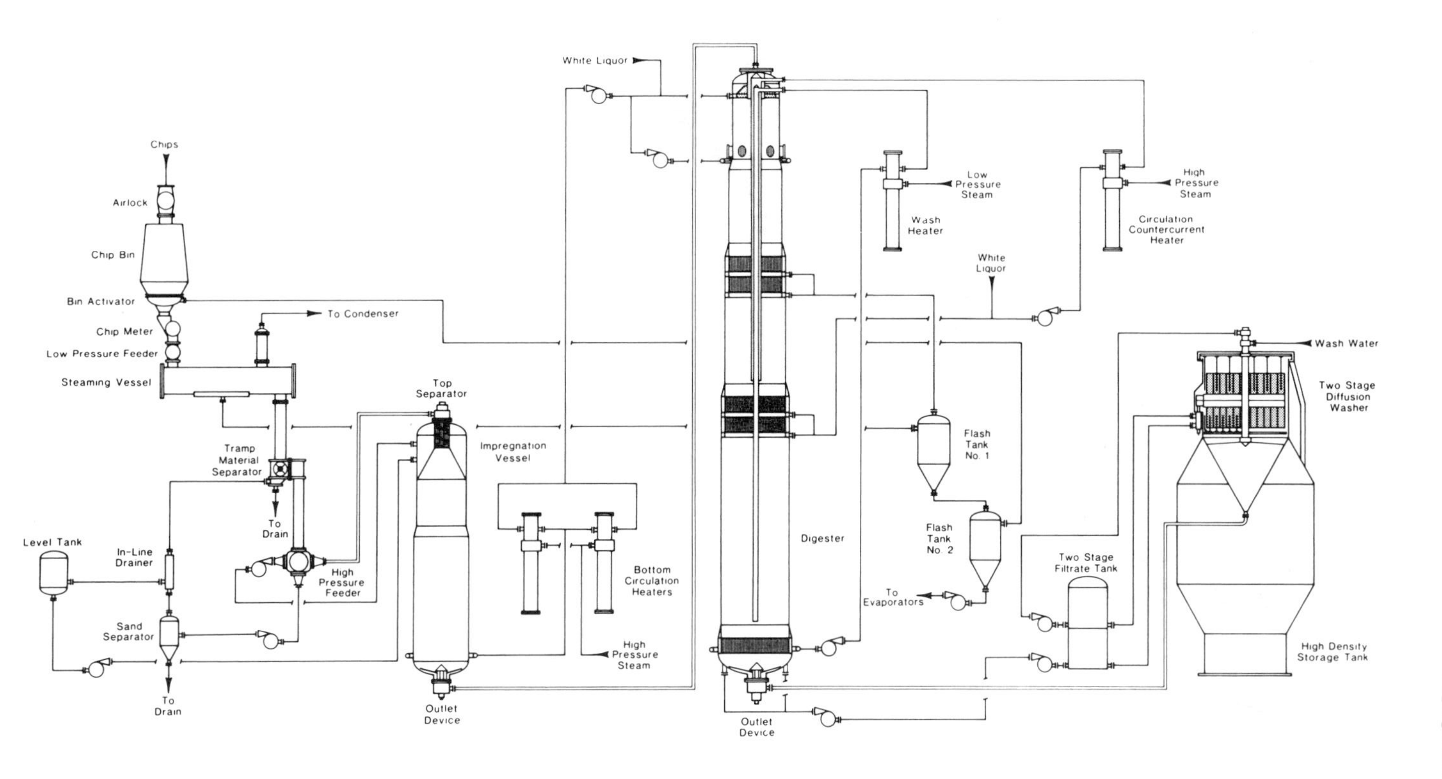

Figure 5.13 Two-vessel hydraulic digester with MCC and two-stage diffuser

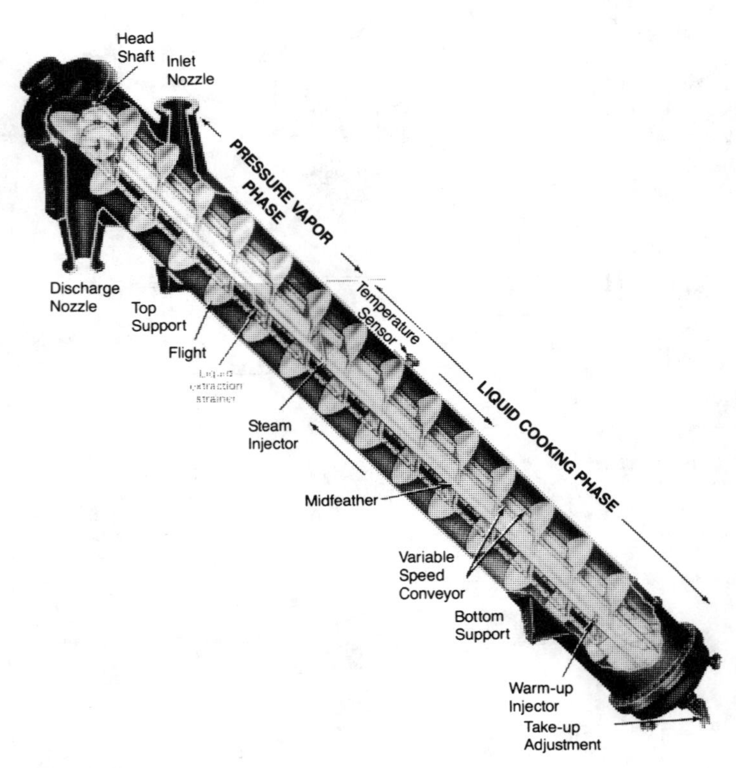

Figure 5.14 M & D digester.

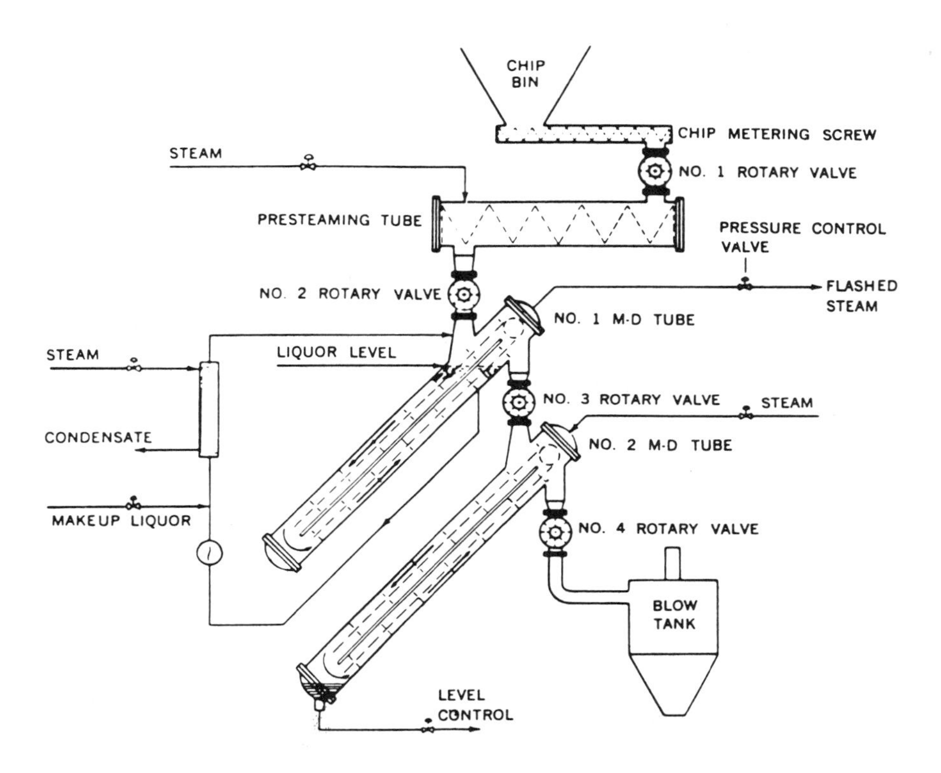

Figure 5.15 Two M & D digesters in series.

Comparison of Batch and Continuous Pulping

Factors favoring batch digesters

Batch digesters are more flexible than continuous digesters. It is easier to change grades, fiber source, and production rate when using a batch digester system. When changing grades or changing from one chip source to another in a continuous digester, some intermediate pulp of inferior quality will always be produced. Because a continuous digester cannot operate too far below its design capacity, production flexibility is limited. When using a set of batch digesters, the production rate can be changed by simply varying the number of digesters in use.

Since several batch digesters are always in use, production reliability is higher than for a continuous digester. When a batch digester breaks down, the production is decreased by maybe 10% to 15%, while breakdown of a single continuous digester means no production at all.

Other advantages of a batch digesting system are that it is easier to start up and shut down than a continuous digester, it gives a higher turpentine yield in by-product recovery, and it requires less maintenance. Batch digesters are also less sensitive to chip fines, which can plug continuous digester screens.

Factors favoring continuous digesters

The main advantages of a continuous digester over conventional batch digesters are the lower steam requirement for heating and the more constant steam demand. This means lower energy costs.

Treatment of non-condensible gases is easier than for batch digesters. A batch digester releases big amounts of gases over a very short time period during digester relief and during blowing. These gases must be contained and taken care of. In continuous digesting these gases are released continuously, making their handling easier.

Another advantage of continuous digesters is that they are more compact and do not require as much space as a batch digester system of the same capacity.

Questions

1. Specify typical dimensions and construction materials for a batch digester.
2. Discuss the different stages of a batch digester cooking cycle.
3. Outline a typical cooking schedule for batch digesting.
4. Describe and compare the different methods for heating a batch digester.
5. Discuss the reason for chip packing.
6. Describe two methods for chip packing.
7. Explain the reason for and procedure of digester relief.
8. Describe the blowing procedure for batch digesters.
9. Explain how a batch digester blow heat system works.
10. Outline the design of a batch digester cold blowing system and describe how it operates.
11. Identify the different pieces of equipment in a Kamyr continuous digester system.
12. Specify typical size and design capacity for a Kamyr digester.
13. Describe the chip and liquor charging system of a Kamyr digester and explain how it operates.
14. Name the different zones in a Kamyr digester.
15. Specify retention times and temperature ranges typical for each zone in a Kamyr digester.
16. Describe the design and explain the function of the different liquor circulation systems in a Kamyr digester.
17. Discuss the washing process in a Kamyr digester washing zone.
18. Recognize what factors will affect the washing efficiency in a Kamyr digester washing zone.
19. Describe the blowing procedure of a Kamyr digester.

20. Explain the difference in design and operation between a steam-liquid phase Kamyr digester and a hydraulic Kamyr digester.
21. Describe the design of the Kamyr two vessel digester system.
22. Name two other continuous digester designs besides the Kamyr digester.
23. List some advantages and disadvantages of the IMPCO digester as compared to the Kamyr design.
24. Describe the design and operation of a M&D digester.
25. Compare the batch and continuous digesting processes and identify factors favoring each process.

Principal Sources For Figures

Ingruber, O.V. "Alkaline Digester Systems," in *Pulp and Paper Manufacture,* Edited by T.M. Grace, B. Leopold, E.W. Malcolm, and M.J. Kocurek. Vol. 5: *Alkaline Pulping.* 3rd ed. Atlanta and Montreal: Joint Textbook Committee of the Paper Industry, 1989.

Ingruber, O.V., Kocurek, M.J., and Wong, A., eds. *Pulp and Paper Manufacture.* Vol. 4: *Sulfite Science and Technology.* 3rd ed. Atlanta and Montreal: Joint Textbook of the Paper Industry, 1985.

Sjodin, L. and Petterson, B. "Two Case Studies on the Cold Blow Technique for Batch Kraft Pulping." *Tappi J.* 70:(2):(February, 1987); 72–76.

Smook, G.A. *Handbook for Pulp and Paper Technologists,* Edited by M.J. Kocurek. Atlanta and Montreal: Joint Textbook Committee of the Paper Industry, 1982.

Fig. 5.1 Smook, Fig. 8.4, pg. 78.
Fig. 5.2 Smook, Fig. 8.5, p. 78.
Fig. 5.5 Smook, Fig. 8.18, pg. 87.
Fig. 5.6 Sjodin, Fig. 4., p. 75.
Fig. 5.7 Ingruber, Fig. 124, p. 157.
Fig. 5.9 Smook, Fig. 8.11, p. 82.
Fig. 5.10 Ingruber, Kocurek, and Wong, Fig. 226, p. 250.
Fig. 5.11 Ingruber, Fig. 129, p. 162.
Fig. 5.12 Ingruber, Fig. 126, p. 159.
Fig. 5.13 Ingruber, Fig. 132, p. 165.
Fig. 5.14 C-E Bauer Brochure
Fig. 5.15 Smook, Fig. 8.9, p. 81.

6
Pulp Processing

Following pulping, the acceptable pulp fibers must be separated from a variety of contaminants. These include uncooked knots and fiber bundles (shives) through screening, pulping liquor through washing, and denser particles through centrifugal cleaning. Different grades of pulp (higher yield, medium yield, bleachable grades) require different processing.

Washing removes pulping liquor by diluting and displacing the dirty black liquor with cleaner wash water. Multiple stages in a countercurrent arrangement provide the highest washing efficiency with the lowest shower water dilution. Operating variables include the pulp itself, consistency, loading level, shower volume and temperature, air entrainment and foam. Washing equipment is represented by cylinder washers, displacement ring washers, pressure diffusers, pressure washers, and belt washers.

Screening removes shives and other oversize rejects. Pressure screens are most commonly used. Screening strategies vary, but the most common arrangement raises the primary screen reject level to maximize efficiency and production. The design of the basket (holes vs. slots) and rotor RPM are key equipment variables.

Centrifugal cleaning removes rejects that have different densities from pulp fibers. Consistency, reject level, and system design affect performance.

Finally, thickening the pulp for storage or market is the final operation found in the pulp mill. From unbleached stage, the pulp is sent to the paper mill or to the bleach plant.

Different Pulp Grades

Introduction

After digesting, the pulp is processed to achieve a homogeneous stock that is clean of black liquor, small wood particles, such as shives, sand, grit, and other dirt.

The pulping process is somewhat different for various kinds of stocks. For example, high-yield pulps must be refined after blowing in order to separate the fibers, while low-yield pulps are cooked long enough for the fibers to be fully separated after blowing, and no extra defibering is needed.

Even though pulp processing depends on the kind of stock being produced, the basic operations are the same. The same kind of equipment is used for pulp washing, screening, cleaning, and thickening no matter what the final product is.

The main differences in the pulp processing operation for different products are:

- Defibering and deknotting operation
- Treatment of rejects
- Targets for screening and cleaning.

High-yield processes

Higher-yield kraft pulp is produced for manufacturing of linerboard, which is a relatively lightweight board used as the outer plies of corrugated boxes. The market for linerboard is very large. In 1987, about 25% of the total paper and paperboard production in the USA consisted of linerboard. Linerboard is always produced in integrated mills; that is, the pulping and papermaking operations are located at the same mill site.

A typical sheet of linerboard consists of two layers: the base, or bottom liner, and the top liner. This two-layer structure is created by using two separate headboxes on the paper machine.

The bottom liner consists of a relatively dark and coarse higher-yield, unbleached pulp, which gives the sheet its stiffness and strength. It is formed from the primary headbox. The top liner is then applied from the secondary headbox. It consists of a lighter, cleaner pulp cooked to a lower yield than the bottom liner. Its function is to completely cover the bottom liner in order to form a good printing surface. The top liner usually makes up 20% to 30% of the total sheet weight. The following shows typical values for yield, kappa number, and percent debris for top and bottom liner.

		Top Liner	Bottom Liner
Yield		49 to 54%	51 to 63%
Kappa		60 to 75	80 to 110
Debris in	batch	10 to 55%	18 to 90%
blown pulp:	continuous	6 to 35%	10 to 90%

The traditional pulping system for linerboard utilizes two separate production lines—one for each liner stock. This is illustrated in Figure 6.1.

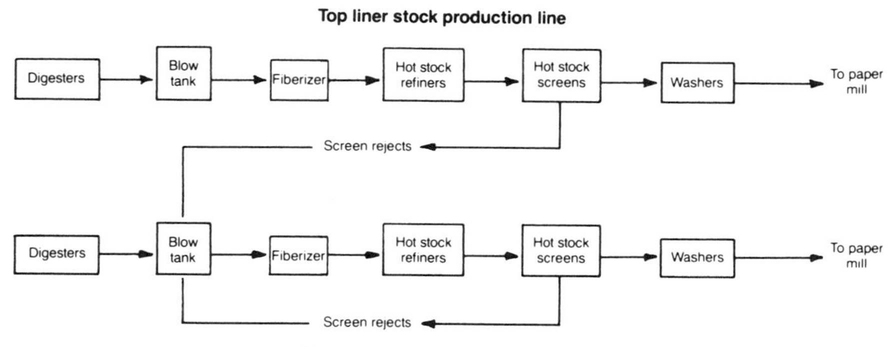

Figure 6.1 Flowsheet for linerboard.

Separate digesters pulp to two separate optimum kappa numbers, and the pulps are defiberized, screened, and washed in separate lines. This system allows maximum control of pulp quality, since the two lines are independent of each other, except for the fact that screen rejects from the top sheet line usually are transferred to the bottom sheet line.

When only one continuous digester is available for pulping, there are different methods for obtaining the right amounts of top and base stock. One such method is to cook for a period using conditions to produce base stock and then switch to conditions for producing top stock. This allows one line of refiners, screens, and washers but has the disadvantage of giving a certain amount of pulp produced at intermediate conditions. Also, there is a limit to what kappa number difference one can achieve between the top and bottom stocks.

Another more acceptable method is to use a stock splitter or Fractionater in the blow line. (See Figure 6.2.) A special screen fractionating system is utilized in order to separate a clean, short-fibered portion of the stock to be used as the top liner stock. The paper mill production lines for top and bottom liner are usually slightly different because of the different properties and requirements of the two.

In the system shown in Figure 6.3, the base stock passes through a fiberizer and hot stock refiners before being washed. After washing, the stock is further refined in a deshive refiner before it is pumped to the paper machine storage tank. The top liner is fiberized and then screened in hot stock screens before washing. The rejects from the screening system are transferred to the base stock line in order to obtain a higher cleanliness for the top liner. After washing, the stock goes through deshive refiners before being pumped to the storage tank.

In linerboard mills, pulp screening is carried out before washing; so-called hot stock screening. The quality requirements for screened pulp vary from mill to mill. Approximate targets are:

Top liner:	0.75% debris
Bottom liner:	1.0% to 1.5% debris

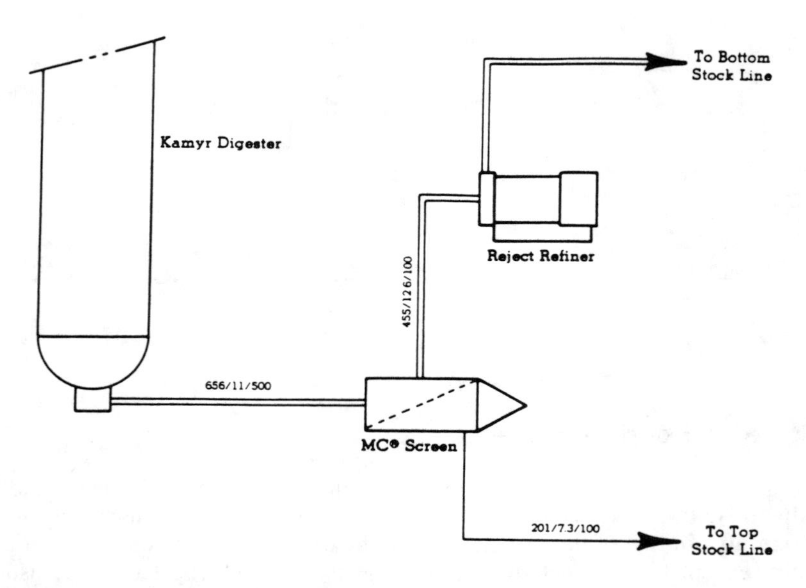

Figure 6.2 Stock splitting after digesting.

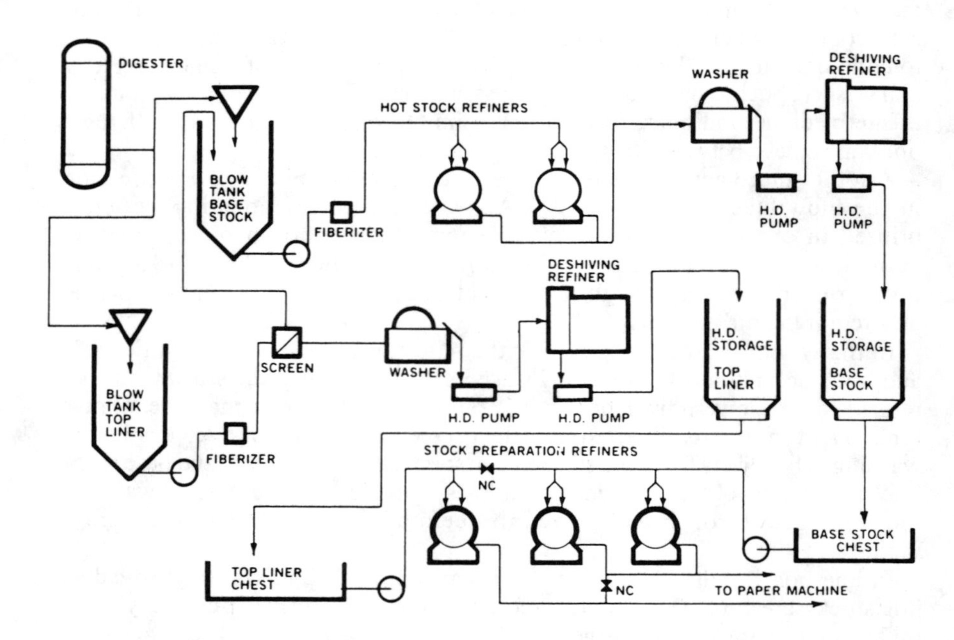

Figure 6.3 Flowsheet with equipment for linerboard.

Medium-yield processes

Medium-yield pulps are used for bag, multi-wall, and saturating grades. These grades are all unbleached. Kappa number is in the range of 35 to 50, yield is 47% to 51%, and percent debris in blown pulp 3% to 28%.

Figure 6.4 shows a standard flow sheet for medium-yield kraft.

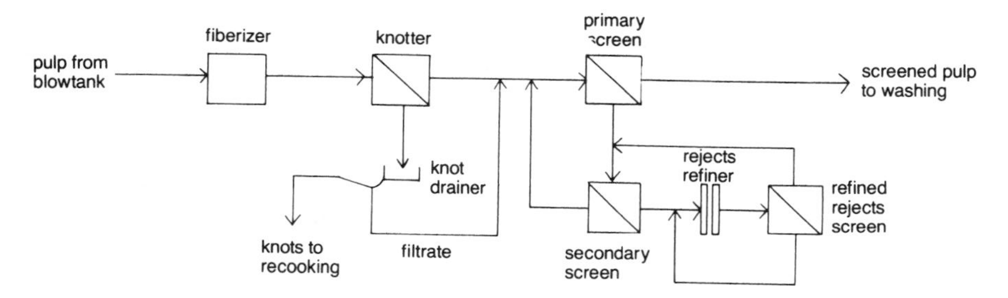

Figure 6.4 Flowsheet for medium-yield kraft

A typical medium-yield pulp processing system includes a fiberizer, knotter, knot drainer, two-stage screening system, and a rejects refiner. This equipment is installed between the blow tank and the washers. A fiberizer, or prebreaker, creates enough mechanical action to break down the cooked chips into fibers. The uncooked, dense knots, which are not fibrilized in the fiberizer, must be separated from the good fibers before screening and cleaning. If not removed, they will interfere with the washing operation, causing lower washing efficiencies and higher soda losses.

Depending on the specifications of the produced grade, the removed knots can be handled in different ways. When the furnish can tolerate darker fibers and higher levels of debris, the knots can be fiberized and returned to the pulp stream. The most common practice, though, is to recook the knots by returning them to the digester. This is not a perfect solution, since dry wood yield from knots is less than 35%, and they take up space in the digester best used for fresh chips.

Two stages of screening are usually enough to achieve desired shive content targets. The rejects from the second stage screen are refined in a disc refiner. A refined rejects screen is often used in order to screen the refined rejects more efficiently than is the case when the refined rejects are simply recycled to the primary screen inlet and mixed with fresh stock.

The rejects from the refined rejects screen are always recirculated to the refiner inlet. The accepts can either be sent to the primary screen feed or to the secondary screen feed. Sending the accepts to the secondary screen feed will result in a very high-system efficiency but also a higher refiner throughput. A third alternative is to send the accepts from the refined rejects screen directly to the system accepts stream. This lowers the overall screening efficiency but also results in a lower refiner loading.

The most common type of screen plate consists of holes. When a very high-quality, medium-yield softwood pulp is required, slotted plates are used—either a single, slotted primary screen or two primary screens where the second screen is slotted.

When screening hardwood medium-yield pulps, slotted screens are usually used in all stages. This is the main difference in treating softwood and hardwood medium-yield pulps.

A centrifugal cleaning system is not normally used for medium-yield grades.

Low-yield processes

Low-yield pulps are used for a range of grades, most of which are bleached to some degree. Some examples of low-yield grades are: bleached paper and board, semi-bleached pulp for newsprint, and dissolving grades.

Bleachable softwood kraft has a kappa number in the range 23 to 35, while kappa for bleachable hardwood is in the range 14 to 20. More recent cooking strategies using extended delignification reduce the kappa number to 12 to 23 for softwood pulps.

For most low-yield pulps a very low shive and dirt content is of utmost importance. Therefore, it is important that knots are removed from the pulp flow as early as possible without breaking down to shives.

In earlier days the pulp was usually washed before it was screened. When screening is done after the last stage of washing, a thickener is required after the screens. This is very costly. By using the decker as the final washer, costs are reduced. If the drained water from the decker is used in the previous stage of cleaning, the system is said to be "closed." When the drained water is sewered, the system is said to be "open." When screening is done between the last two stages of washing, it is called inter-stage screening. (See Figure 6.5).

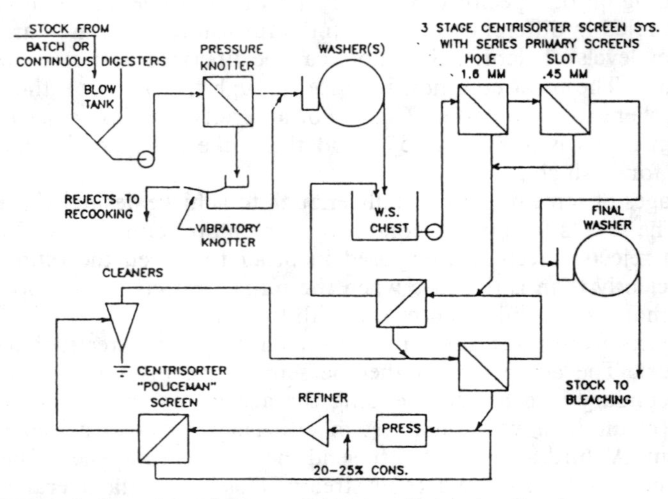

Figure 6.5 Basic flowsheet for low-yield bleachable kraft pulp.

Today it is possible to install the knotting and screening systems immediately after the blow tank and before the washing system. This can be done because pressure knotters and pressure screens allow the screening to take place without air entrainment and foaming. New screening systems are usually installed before washing even in existing mills, unless there are constraints making this impossible.

The equipment used in a low-yield pulp processing system is basically the same as for medium-yield pulps. The size of the perforations in knotter and screens is normally smaller, however.

For bleachable grades, three or four stages of screening is common. When the rejects are sewered, as often is the case with high-quality, bleachable grades, a higher fiber recovery is achieved by using an extra stage of screening.

The handling of screen rejects varies depending on what grade is being produced. Rejects can be thickened and removed from the system, which always produces the highest quality of pulp, or they can be refined and kept in the system. Rejects removed from the system can be sent to landfill, burned in a hog fuel boiler, recooked in a batch digester, or sent to a high-yield line. Recooking may not be the best process option, but it does reduce soda losses, as compared to dumping the rejects.

When the screen rejects are refined they are usually returned to the last stage screen feed. Some systems have a refined rejects screen, then the accepts are returned to the primary screen feed.

Cleaners are usually installed in the rejects loop to remove sand and grit, which would otherwise build up in the system.

When the rejects are removed from the system, it is common to use a slotted plate scalping screen in the last stage. This kind of screen is more efficient than a pressure screen for removing the remaining good fibers from the rejects. Before the rejects are dumped, they are drained to recover the liquor.

Defibering and Deknotting

Objective of defibering

High- and medium-yield pulps do not disintegrate fully into separate fibers during blowing. Some additional mechanical action is, therefore, required to complete the disintegration of the cooked chips into fibers.

There are two types of equipment used for defibering: *prebreakers* and *refiners*. Other names for the prebreaker are fiberizer or fibrilizer. The prebreaker gives the pulp a milder mechanical treatment than the refiner and is used for defibering medium-yield pulps before screening and washing. Prebreakers are also installed before the hot stock refiners in high-yield pulp production lines. Refiners are used for hot stock refining of high-yield pulps before washing. The bulk of the defibering should be done before washing, since the washing efficiency is low for pulp containing large fiber bundles.

For pulps with very high kappa numbers (above 80), so-called deshive refining is required. It should be carried out at a high consistency in order to avoid fiber shortening and loss in tear strength. Therefore, deshive refining is often done after the washers, where the pulp is discharged at a consistency of 10% to 15%.

Rejects from knotters and screens can also be refined and returned to the pulp flow. Rejects refining is not done for all pulp grades, however.

Prebreakers and refiners

A modern prebreaker or fibrilizer operates on the principle of hydraulic shear and friction and wear of large fiber bundles against each other. Figure 6.6 shows a prebreaker. Most designs contain some kind of rotor and a perforated plate, through which the fiberized pulp must pass.

Figure 6.6 Prebreaker.

Refining of pulp and rejects is carried out in some type of disc refiner. The refiners must be sealed so air cannot enter the system. A typical disc refiner consists of a double-sided rotating disc located between two stationary discs. The pulp is fed to the center of the refiner and then flows radially outward between the rotating disc and the stationary discs. The refined pulp is collected in the refiner casing and is discharged. (See Figure 6.7).

Objective of deknotting

Knots consist of uncooked wood particles. A general definition of knots is the fraction of pulp that is retained on a screen with 9 mm holes. These rejects can be "true knots," i.e., particles of compression wood and resinous heartwood, or undercooked particles remaining in the pulp due to overthick chips.

The knots should be removed as early in the process as possible, before pulp washing, since their presence would decrease the washing efficiency. In high-yield kraft pulps the knots are broken down to smaller particles by prebreaking and refining. No special knotter screen is required then, since the

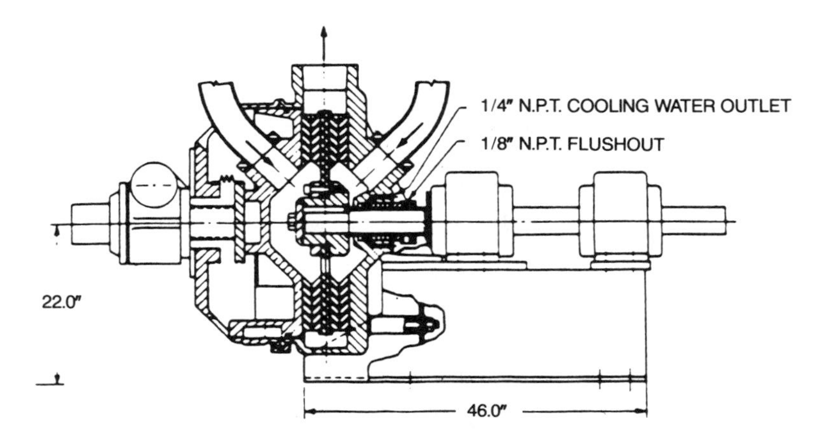

Figure 6.7 Disc refiner.

particle size is reduced enough in the refiner. Medium-yield systems generally utilize a prebreaker followed by a knotter while low-yield systems use a knotter only. If a low-shive content is desired, a prebreaker is not utilized, since broken knots from prebreaking can cause an increase in pulp shive content.

Knotter designs

Two major types of knotters are in current use: vibrating screen knotters and pressure knotters.

The vibrating screen knotter is the older design, which is rapidly being replaced by the newer pressure knotter. It has many disadvantages, such as problems with foam and introducing air into the stock, which will decrease the washing efficiency.

Pressure knotters are clean, require no hood, and do not introduce any air into the stock. (See Figure 6.8.) The operation of a pressure knotter is basically the same as for a pressure screen. The main difference is that knots are much easier to separate from the stock than shives, due to their size and rigidity. The main drawback of the pressure knotter is that good fibers are rejected together with the knots. A second screen is needed for fiber recovery. Here, the vibrating screen still comes to use. Good fibers are recovered and liquor is drained from the knots on this screen, called the *knot drainer.*

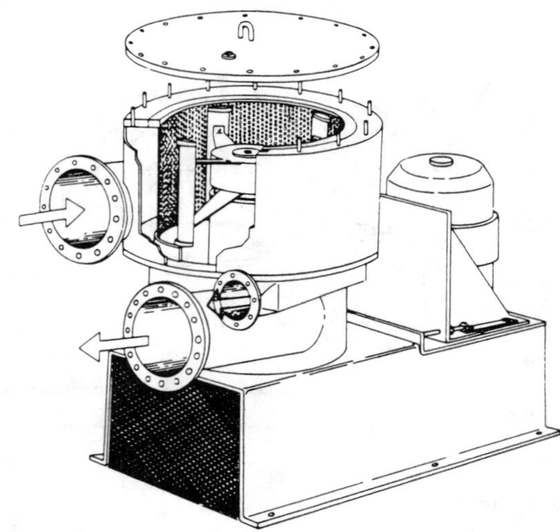

Figure 6.8 Pressure knotter.

Brown Stock Washing

Objective of brown stock washing

Brown stock washing differs from digester washing. In a continuous digester zone, cooked chips not yet disintegrated into individual fibers are washed; whereas, in external washers, pulp is washed.

The objective of brown stock washing is to remove the maximum amount of black liquor dissolved solids from the pulp while using as little wash water as possible.

The dissolved solids left in the pulp after washing will interfere with later bleaching and papermaking and will increase costs for these processes. The loss of black liquor solids due to solids left in the pulp means that less heat can be recovered in the recovery furnace. Also, makeup chemicals must be added to the liquor system to account for lost sodium compounds.

It would be easy to achieve very high washing efficiencies if one could use unlimited amounts of wash water. As it is, one has to compromise between high washing efficiency and a low amount of added wash water.

The water added to the liquor during washing must be removed in the evaporators prior to burning the liquor in the recovery furnace. This is a costly process and often the bottleneck in pulp mill operations. Minimizing the use of wash water will therefore decrease the steam cost for evaporation.

Basic pulp washing mechanisms and operations

There are two separate mechanisms involved in any washing operation—washing and leaching.

Washing is simply bulk removal of the liquor surrounding the pulp fibers. This is done by removing the stronger liquor and replacing it with weaker liquor or water.

Leaching includes diffusion and desorption. Dissolved solids and chemicals in the interior of the pulp fibers must diffuse into the surrounding liquor before they can be removed. The diffusion rate of sodium is quite rapid while diffusion of larger molecules, such as dissolved lignin, is much slower.

Sodium and dissolved solids are also attached or bound to the pulp fiber surfaces to a certain degree. Desorption occurs when these attached compounds leave the fiber surface for the surrounding liquor. Desorption and diffusion are most significant in the final washing stages when the liquor surrounding the pulp fibers has a low concentration of dissolved solids and chemicals.

The two basic pulp washing operations are dilution/extraction and displacement washing.

In dilution/extraction washing, the pulp slurry is diluted and mixed with weak wash liquor or clean water. Then the liquor is extracted by thickening the pulp, either by filtering or by pressing. This procedure must be repeated many times in order to sufficiently wash the pulp. The efficiency of dilution/extraction washing depends primarily on the consistencies to which the pulp is diluted and thickened. (See Figure 6.9.)

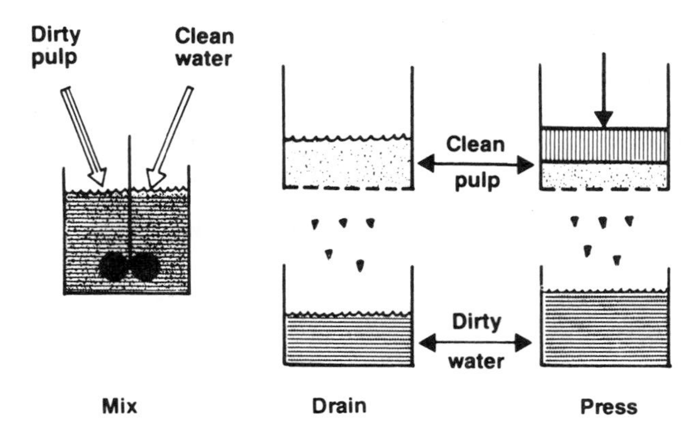

Figure 6.9 Principle of dilution/extraction.

In displacement washing, the liquor in the pulp is displaced with weaker wash liquor or clean water. Ideally, no mixing takes place at the interface of the two liquors. In practice, however, it is impossible to avoid a certain degree of mixing. Some of the original liquor will remain with the pulp and some of the wash liquor will channel through the pulp mass. The efficiency of displacement washing then depends on this degree of mixing and also on the rate of desorption and diffusion of dissolved solids and chemicals from the pulp fibers. (See Figure 6.10.)

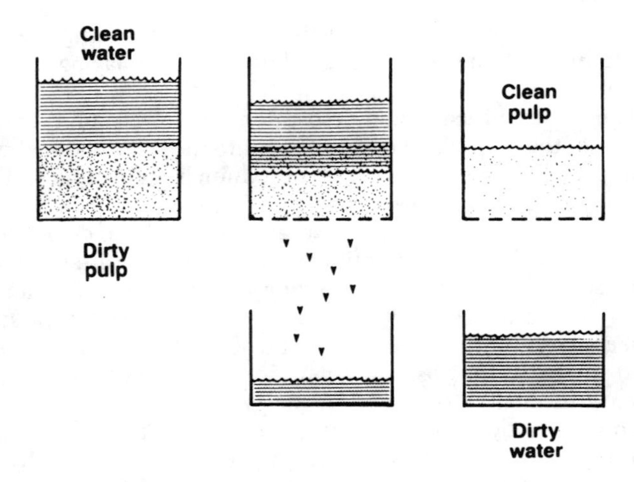

Figure 6.10 Principle of displacement washing.

All pulp washing equipment is based on one or both of these basic principles. Displacement washing is utilized in a digester washing zone. A rotary vacuum washer utilizes both dilution/extraction and displacement washing, while a series of wash presses utilizes dilution/extraction.

Most pulp washing systems consist of more than one washing stage. The highest washing efficiency would be achieved if fresh water were applied in each stage. However, this approach would require large quantities of water and is, therefore, not used.

Countercurrent washing is the generally-used system design. In countercurrent washing, the pulp in the final stage is washed with the cleanest available wash water or fresh water before leaving the system. The drained water from this stage is then sent backwards through each of the previous stages in a direction opposite to the pulp flow.

Terminology used in washing

Soda loss

Most pulp mills still express the efficiency of their washing systems in terms of "soda loss." *Soda loss* is defined as the sodium content of the pulp leaving the washing system, expressed as pounds of salt cake per ton of oven-dry pulp. "Salt cake" is another name for sodium sulfate.

Actually, very little of the sodium in the pulp is in the form of sodium sulfate. Soda loss was originally expressed in this manner because the sodium lost with the pulp was always replaced with salt cake added to the dissolving tank.

The ratio of sodium to organic compounds in the dissolved solids can vary widely and is dependent on factors such as wood species used and pulp product produced. This ratio also changes near the end of the washing operation.

Sodium diffuses out from the cell wall more easily than the bigger lignin fragments and so a higher proportion of sodium is removed from the pulp than organic material.

An alternate method of reporting washing efficiency would, therefore, be to determine and report the loss of total dissolved solids per unit of pulp, then convert to sodium equivalents. Each mill would need to determine a set of correlations between sodium loss and dissolved solids loss specific for the different conditions encountered at that mill.

When reporting washing efficiency in relation to salt cake loss, it is important to indicate whether it is total soda loss or washable soda loss that is reported. The sodium in the pulp consists of two fractions, washable sodium, which is water soluble and can be removed to 100% by thorough washing, and sorbed sodium, which consists of sodium bound to the acidic groups in the kraft pulp. The level of sorbed sodium varies with kappa number and pH. (See Figure 6.11.)

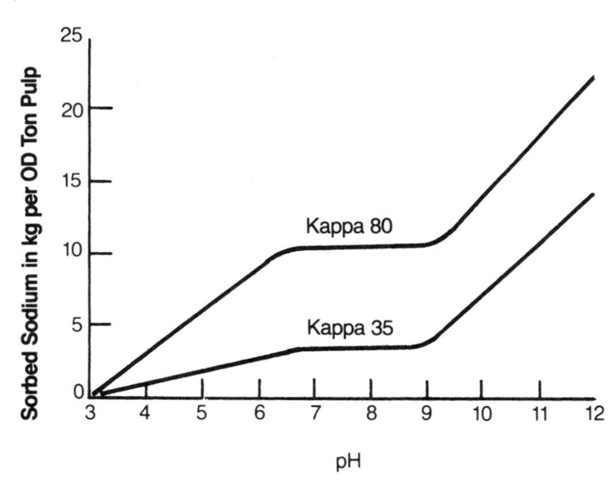

Figure 6.11 Effect of pH on sorbed sodium.

The total soda loss, therefore, will not tell one very much about the actual washing efficiency. It is an important factor in determining the amount of makeup chemical required, however.

Dilution factor

The quantity of water used for washing pulp is normally expressed as *dilution factor.* It can be calculated for a single washing stage or for a whole system and is adaptable to all types of washing. Dilution factor is defined as the weight of wash water introduced into the black liquor per unit weight of oven-dry pulp being washed. The wash water added to the black liquor is the difference between wash water applied to the washing unit and the water remaining in the pulp after washing. A negative dilution factor means that less wash water is added than remains with the pulp after washing.

The dilution factor can be expressed mathematically as:

$$DF = \frac{F}{P} - \frac{100 - C}{C}$$

F = wash water flow to the stage (tons/hr)
P = pulp throughput (tons/hr)
C = consistency of pulp leaving stage (%)

A higher dilution factor means more wash water added to the system, which results in a higher washing efficiency. The economical benefits of an increased washing efficiency must be balanced against the increased costs of evaporation, however. Dilution factors are normally in the range of *2.0 to 3.0.*

Displacement Ratio

Displacement Ratio (DR) is a term that expresses the effectiveness of a single displacement washing stage in removing solids from the pulp. It is defined as the ratio of the actual reduction in solids content in the stage compared to the maximum possible reduction.

$$\text{Displacement Ratio} = \frac{C_v - C_s}{C_v - C_w}$$

Cv = concentration of dissolved solids in the liquor entering the stage with the pulp. (For rotary vacuum washers: Cv = concentration of dissolved solids in the washer vat.)
Cs = concentration of dissolved solids in the liquor leaving the stage with the pulp.
Cw = concentration of dissolved solids in the wash water added to the stage.

In the ideal case, the liquor in the pulp leaving the washer would have the same concentration of dissolved solids as the wash water added, i.e., DR = 1.0. In reality, however, displacement ratios for rotary vacuum washers are in the range of 0.60 to 0.90.

Norden Efficiency Factor

The Norden method is a way of comparing the efficiency of different kinds of washing equipment independently of dilution factor. The method assumes that a washing stage can be likened to a number of countercurrent mixing stages connected in series. Pulp slurry, which enters one stage, is mixed with liquor from the next stage. The stock is then rethickened to the original consistency, and the separated pulp and liquor pass on to the next stages.

The original method of calculating was slightly modified. The modified Norden Efficiency factor, N, is now defined as the number of mixing stages that will give the same results as the washing equipment under consideration when operated at a standard consistency and at the same dilution factor as the equipment in question. (See Figure 6.12.)

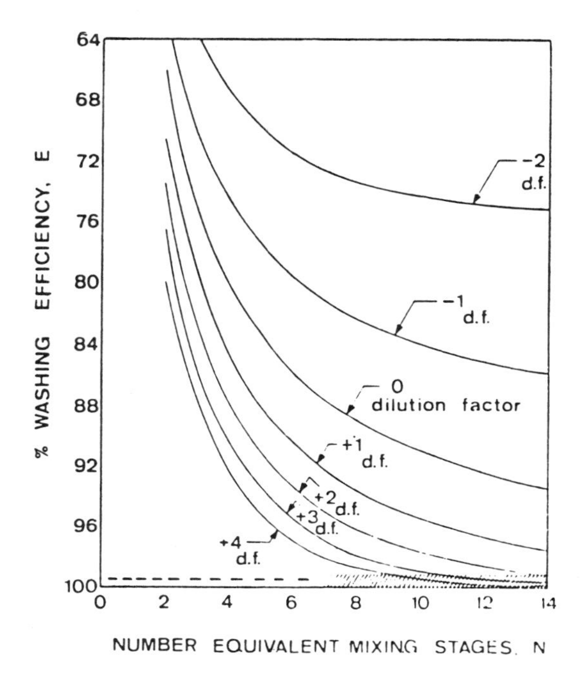

Figure 6.12 Modified Norden Efficiency Factors.

Washing equipment

There is a large variety of pulp washing equipment available these days. The rotary vacuum washer is still by far the most widely used washer equipment, but other methods of washing have been developed and are becoming more common. Among these are rotary pressure washers, diffusion washers, horizontal belt filters and wash presses.

Rotary vacuum washers

The rotary vacuum washer consists of a wire- or cloth-covered cylinder that rotates in a vat containing the pulp slurry. A schematic of a rotary vacuum washer is shown in Figure 6.13. Vacuum is applied from the inside of the cylinder and a pulp mat is formed on the surface of the cylinder when in the vat. As the cylinder continues to rotate, the pulp mat emerges from the vat, and wash water is applied with showers. The mat is continuously dewatered by the vacuum applied.

Finally, the vacuum is cut off, and the washed pulp mat is removed from the cylinder. In a multistage system, the washed pulp mat is diluted with new wash liquor and is transported to the next washer vat. From there, the whole washing process is repeated. Consistency in the washer vat is normally around one percent, while consistency of the pulp mat leaving the washer is between 9% to 18%.

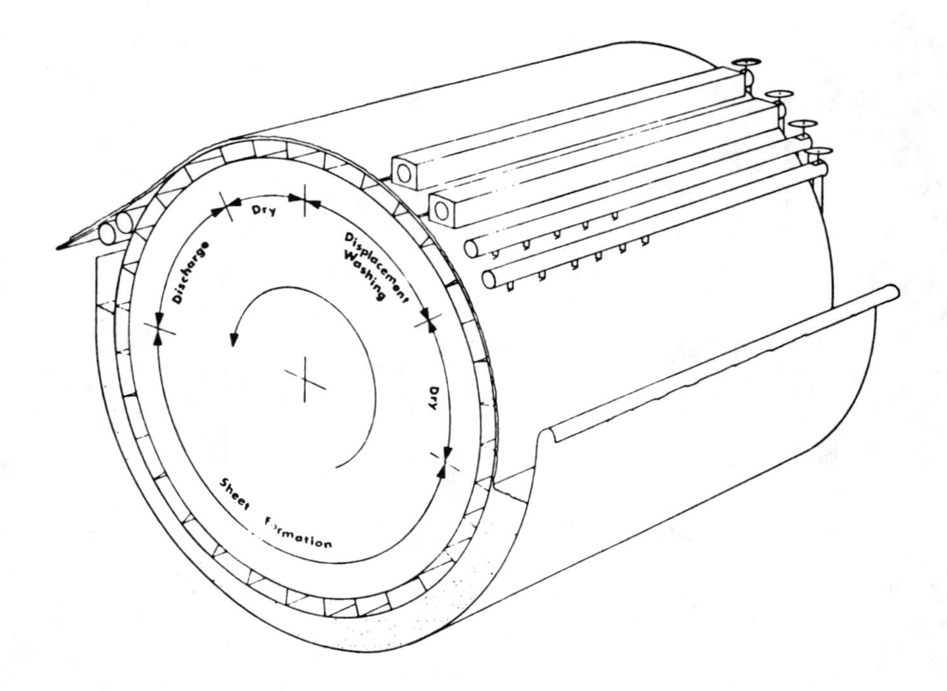

Figure 6.13 Schematic of rotary vacuum washer.

A rotary vacuum washer uses both the dilution/extraction and the displacement washing principles. The pulp is diluted in the washer vat, liquor is extracted from the pulp mat by the applied vacuum, and remaining liquor in the pulp mat is displaced with wash water from the showers.

The liquor extracted from the pulp mat is collected on the inside of the drum and is carried down the drop leg, or barometric leg, into the seal tank. It is this drop leg that creates the necessary vacuum inside the drum. A drop leg height of about 8.5 m to 11 m is necessary to create operating vacuums of 18 cm to 30 cm of mercury. Figure 6.14 shows a drop leg configuration.

Some air is always pulled through the pulp mat and gets mixed with the extracted liquor. This is impossible to avoid. The air/liquor mixture generates foam. It is important that the seal tank is large enough to allow entrained air bubbles to escape from the liquor before it is pumped to the next washing stage. In a multistage washing system, the top sections of all seal tanks are generally connected to a common foam tank to which excess foam escapes and settles.

A multistage rotary vacuum washer system usually contains two to five units. If the pulp is digested in a continuous digester with a washing zone, one or two stages of drum washing are enough. Three to five stages are required if no previous washing occurred. The wash water and the pulp flow in opposing

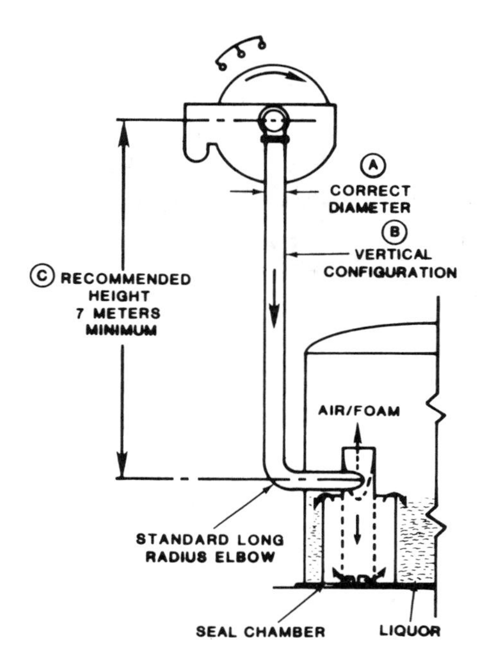

Figure 6.14 Drop leg configuration.

directions. This is called countercurrent washing. Figure 6.15 shows a countercurrent 3-stage rotary vacuum washer system.

The cleanest wash water is applied at the showers of the last stage washer. The filtrate from the drum of this last stage is then used to dilute the pulp entering this stage and in the showers of the previous washer. The wash liquor gets stronger for each stage and, in the first stage washer, the strongest wash liquor washes the dirtiest pulp. After the liquor has been extracted from the first stage washer, it is pumped to the recovery system. If there is a digester washing zone, however, this liquor is used in the digester washing zone.

The most important operating variable for rotary vacuum washers is the *dilution factor.* Some other variables affecting the washer operation are:

- Specific loading
- Vat consistency
- Shower flow distribution
- Wash water temperature
- Drum speed
- Amount of air in stock
- Type of stock.

Specific loading is defined as oven-dry (o.d.) tons per day of pulp per square foot of a single stage cylinder surface. Typical design specific loadings are in the range of 0.6 to 0.8 oven-dry tons/day/square foot. A vacuum washer

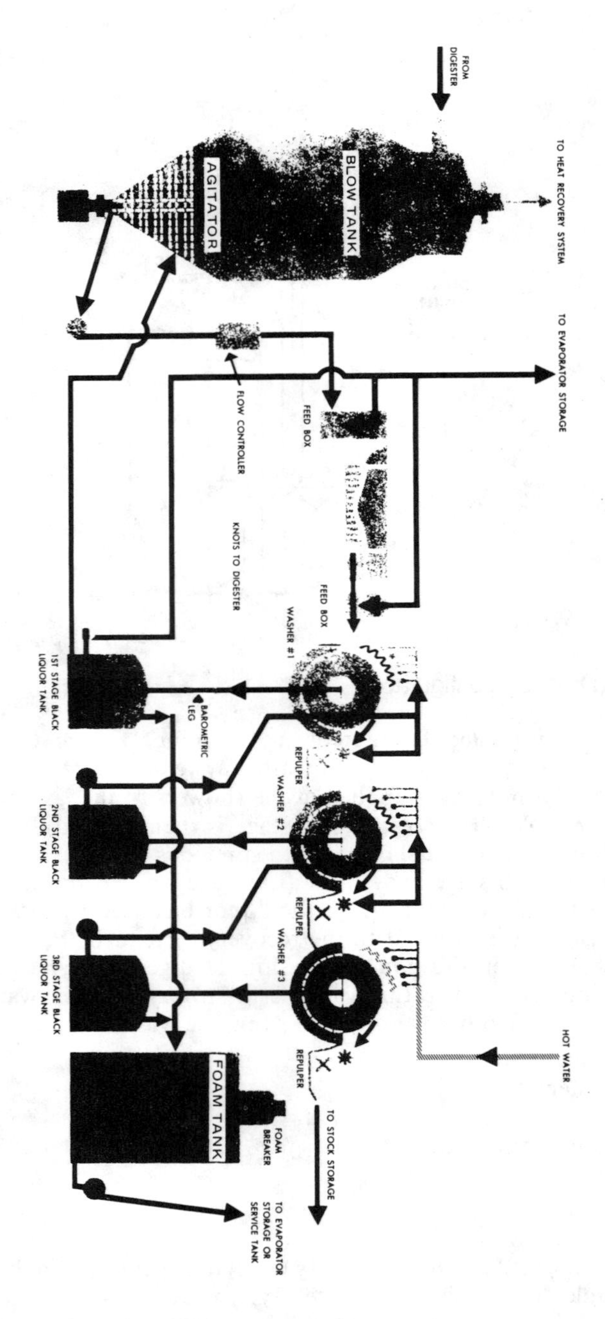

Figure 6.15 Washing system including tanks.

system is flexible and can operate at specific loadings much higher than what they were designed for, but the overall washing efficiency will decrease correspondingly.

The vat consistency must not be too high in order for a well-formed, easily-drained pulp mat to form on the drum. It is also more difficult to achieve proper mixing with the dilution water when working at higher consistencies.

An adequate shower flow distribution is important to achieve a high displacement ratio and avoid mixing and channeling. The temperature of ingoing wash water should be in the range of 50 °C to 80 °C. Washing is poor at cool temperatures and at hot water temperatures above 80 °C.

Drum speed is a variable that is dependent on vat consistency and/or pulp throughput. A high drum speed combined with a low-vat consistency results in a thinner pulp mat and gives better washing results than a low drum speed combined with a high-vat consistency, resulting in a thick, unevenly formed pulp mat. An optimum speed-consistency relationship results in a high-vat consistency improving the displacement ratio.

Air in the stock has an adverse effect on pulp washing, since it lowers the drainage rate of the pulp mat, thus lowering capacity. Therefore, it is important to separate air entrained in the washer filtrate before it is used again.

The type of stock has a big impact on obtainable overall washing efficiency. Hardwood pulps are, for example, easier to wash than softwood pulps due to the greater foaming tendency of softwood liquor. Pulps of high kappa numbers are more difficult to wash. They contain fiber bundles holding black liquor solids that will not have enough time to fully diffuse into the surrounding liquor during the washing process.

Rotary pressure washers

A rotary pressure washer is very similar to a rotary vacuum washer. The main difference is that an external pressure is used for mat formation and dewatering instead of an internal vacuum. A pressure washer is shown in Figure 6.16. The cylinder and vat are fully enclosed in a washer hood in order to be able to raise the pressure in the air space surrounding the drum.

The main difference in drum construction is the absence of channels on the inside. The interior of the washer drum can, therefore, be utilized for a more sophisticated liquor collection system. The washer can be operated with two or three displacement stages by having separate draining systems for each set of showers. The cleanest filtrate, collected under the last stage of showers, is led back to the previous set of showers on the same washer. The most concentrated filtrate is used for dilution of the pulp before the washer and for the cleanest showers on the preceding washer, (see Figure 6.17), a typical countercurrent operation.

Pressure washers offer a few advantages compared to vacuum washers in addition to having up to three displacement stages. The higher pressure reduces foaming and allows higher temperature wash liquor to be used. The closed vapor circulation system makes it possible to collect and treat odorous vapors released during washing.

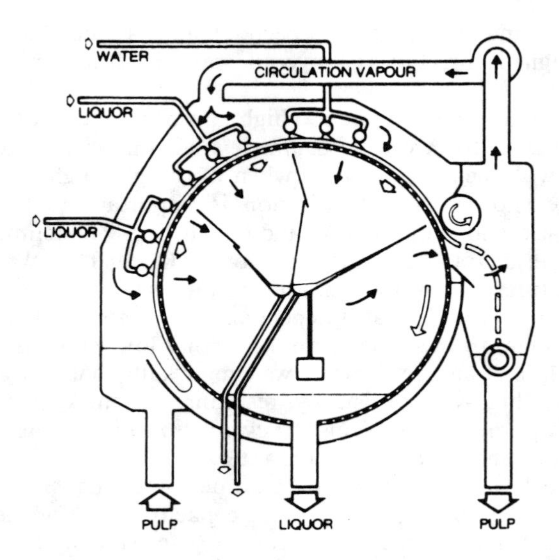

Figure 6.16 Principle of a pressure washer.

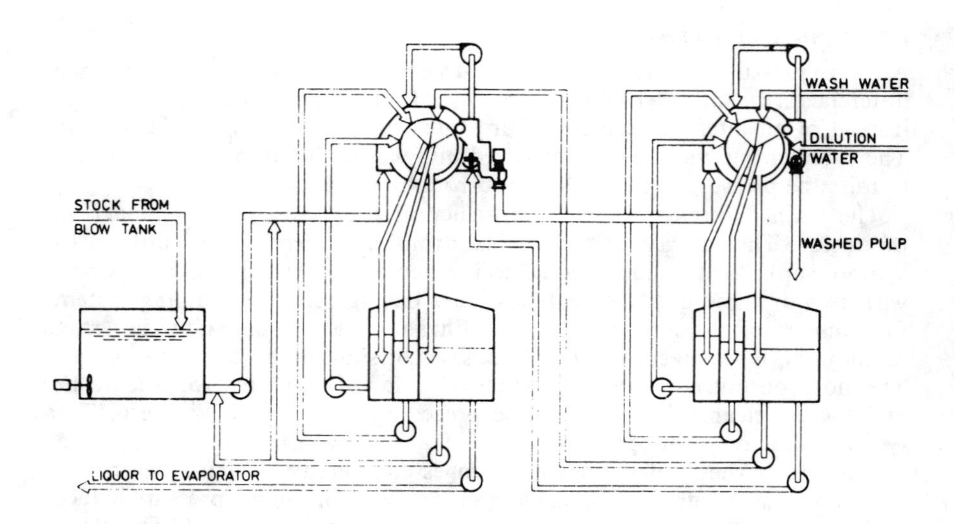

Figure 6:17 Flowsheet for a two-stage pressure washer system.

Diffusion washers

Diffusion washers are pure displacement washers. There are two types: one stage or multistage ring diffusion washers, and pressure diffusion washers.

The *ring diffusion washer* unit consists of a series of screen rings, one inside another. (See Figure 6.18.) Each ring is hollow and has perforations on both sides so liquor can be drained through it. Each ring is connected to a radial drainage arm through which the drained liquor is removed. The whole ring assembly is mounted to a set of hydraulic cylinders and can be moved up and down. Figure 6.19 shows a close-up of one ring.

The pulp enters the diffusion washer at the bottom and moves upward through the annular spaces between the screen rings. The wash liquor is introduced into the pulp through a set of nozzles, mounted on a rotating radial arm. Each nozzle is located at the mid-point between two screen rings. As the nozzle arm rotates, it leaves a string of wash liquor behind. This wash liquor then moves radially through the pulp in both directions toward the screen rings, thus replacing the liquor entering with the pulp. The displaced liquor is removed through the drainage arm and flows to a storage tank.

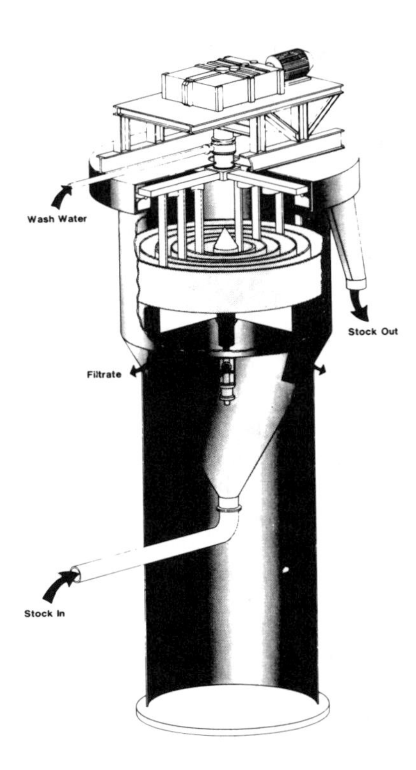

Figure 6.18 Ring diffusion washer.

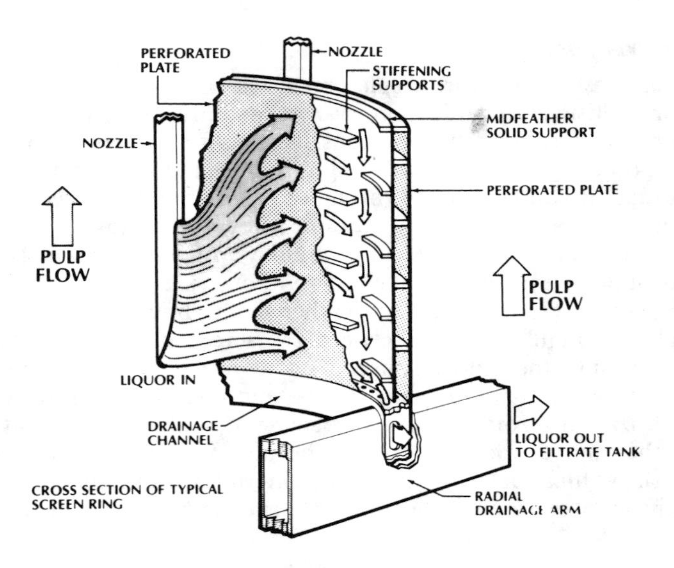

Figure 6.19 Flows in a ring diffusion washer.

The screen rings move upward with the same velocity as the upward flowing pulp. When the screen ring assembly has reached its top position, it is moved rapidly downward in order to keep the screens clean and prevent them from plugging.

In a multistage diffusion washer the units are mounted on top of each other. They work in countercurrent fashion; that is, displaced liquor from one stage is used as wash liquor in the preceding stage. A multistage ring diffusion washer is shown in Figure 6.20.

Typical retention times for one stage in a ring diffusion washer is eight to ten minutes. This gives plenty of time for diffusion of liquor from the interior of the fibers.

The *pressure diffusion washer* is built to operate at digester pressure and is connected to the blow-line of a continuous digester. The washer consists of a pressure vessel with a central body through which the wash liquor is introduced. A moving cylindrical screen is located at the periphery of the vessel. Figure 6.21 shows a pressure diffusion washer.

The pulp is introduced at the bottom of the washer and flows upwards in the annular space between the screen and the outer shell. Wash liquor is injected into the pulp from nozzles located in the outer shell. It flows radially through the pulp bed, displacing the liquor surrounding the pulp. The displaced liquor is extracted through the screen and is collected in the central collection chamber. The cylindrical screen moves upward with the pulp. A periodic fast downstroke keeps the screen clean.

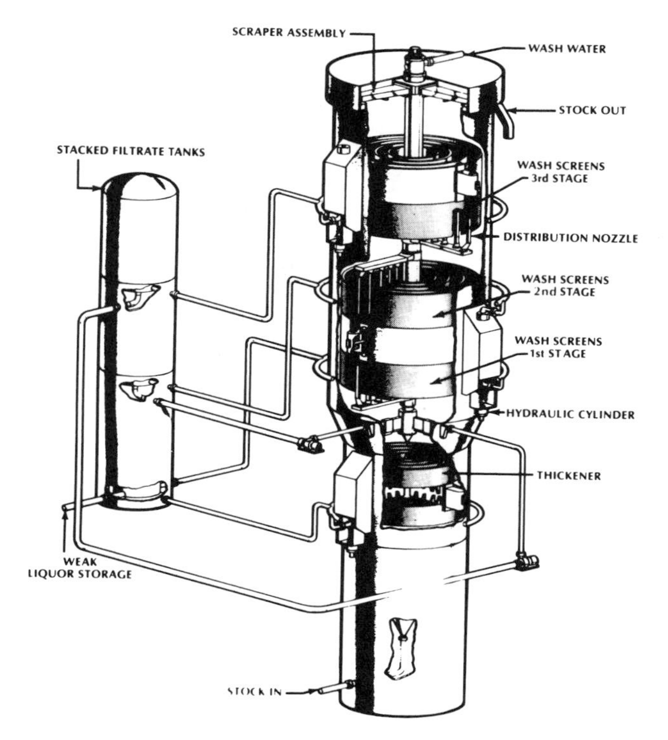

Figure 6.20 Multistage ring diffusion washer.

Diffusion washers have some advantages compared to rotary vacuum washers:

- Diffusion washing takes place in a submerged environment, which excludes the possibility of air entrainment and foaming.
- There is no release of odorous gases.
- They take up less space.
- They are easy to operate and require little process control instrumentation.

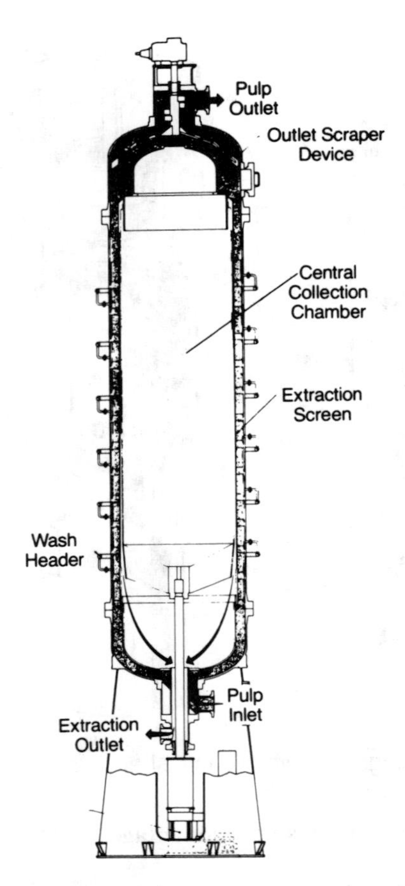

Figure 6.21 Pressure diffusion washer.

Horizontal belt washers

The belt washer is a new kind of pulp washing equipment that resembles the fourdrinier section on a paper machine. (See Figure 6.22.) Pulp slurry at a consistency of up to 3.5% is distributed across a traveling wire from a headbox. The pulp is dewatered, forming a mat of about 8% to 12% consistency. This pulp mat is then washed in a series of displacement stages as it moves from the headbox to the couch roll. The mat's consistency remains relatively constant at 8% to 12% during these washing stages.

The cleanest wash water is added in the final shower ahead of the couch roll. It is drained through the mat with a suction box, then sent to the previous shower. The filtrate from the first shower is finally sent to the evaporators, while the filtrate drained ahead of the first shower is used for dilution of the headbox furnish.

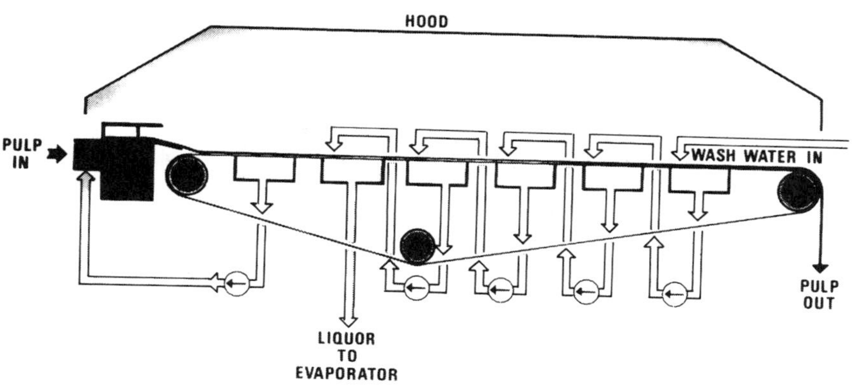

Figure 6.22 Horizontal belt washer.

Belt washers, thus, use one dilution/extraction stage, followed by several displacement stages. Theoretically, a large number of displacement stages can be fit along the wire. Belt washers are supposed to give high overall washing efficiencies at comparatively low dilution factors.

The wire on the belt washer can be of three basic designs. It can be a grooved rubber belt, a woven plastic filament belt, or made of a thin sheet of solid stainless steel, which has been perforated and welded to become a continuous belt.

Wash presses

Basically, wash presses are dilution/extraction washers, although some wash presses also have a displacement stage. Wash presses are frequently used for pulps that are difficult to permeate.

In order to reach sufficient efficiencies, a press must be able to reach a very high discharge consistency up to 40% or 50%. A wash press system consists of a number of presses arranged in a series with an agitated tank for dilution between them. (See Figure 6.23.) A system's efficiency mainly depends on two factors: the degree of equilibrium reached in the agitation tank and the degree of extraction in the presses.

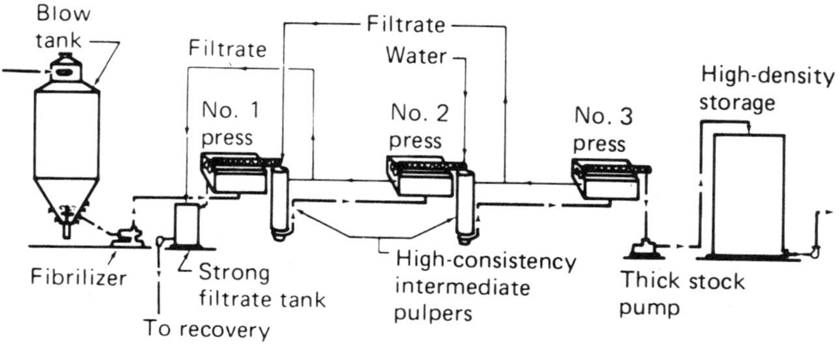

Figure 6.23 Wash press system.

There are a number of different press designs. Figure 6.24 shows a screw press and Figure 6.25 shows a twin-roll press.

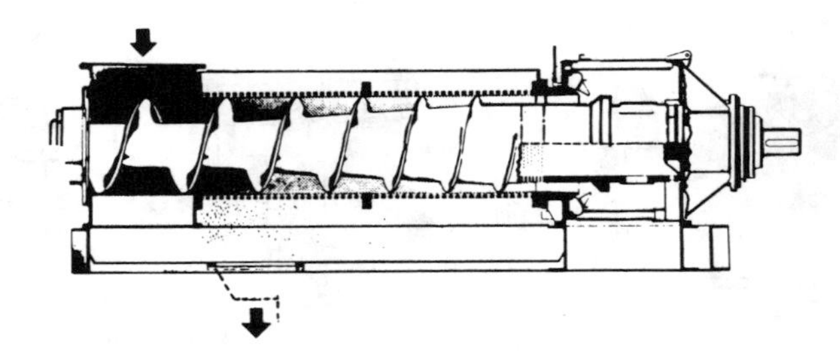

Figure 6.24 Screw press.

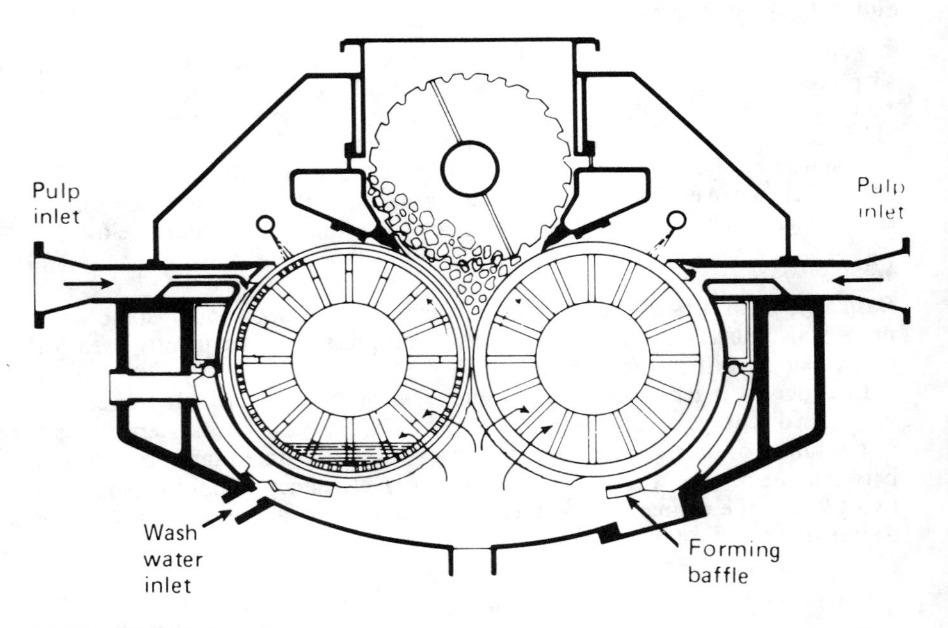

Figure 6.25 Twin-roll press unit.

Washing in continuous digesters

As was seen in Chapter 5, the washing that takes place in a continuous digester's washing zone is called diffusion washing. It is pure countercurrent displacement washing. The cooked chips flow downward, while the wash liquor, introduced at the bottom of the digester, flows upward and is finally extracted along with the mother liquor through the extraction screens.

The retention time of the chips in the washing zone is normally from 1.5 to 4 hours, depending on what washing efficiency the digester was designed to achieve. A long retention time is required in order for the dissolved solids entrained inside the chips to have enough time to diffuse into the surrounding liquor. The washing efficiency is a function of retention time and dilution factor as can be seen in Figure 6.26.

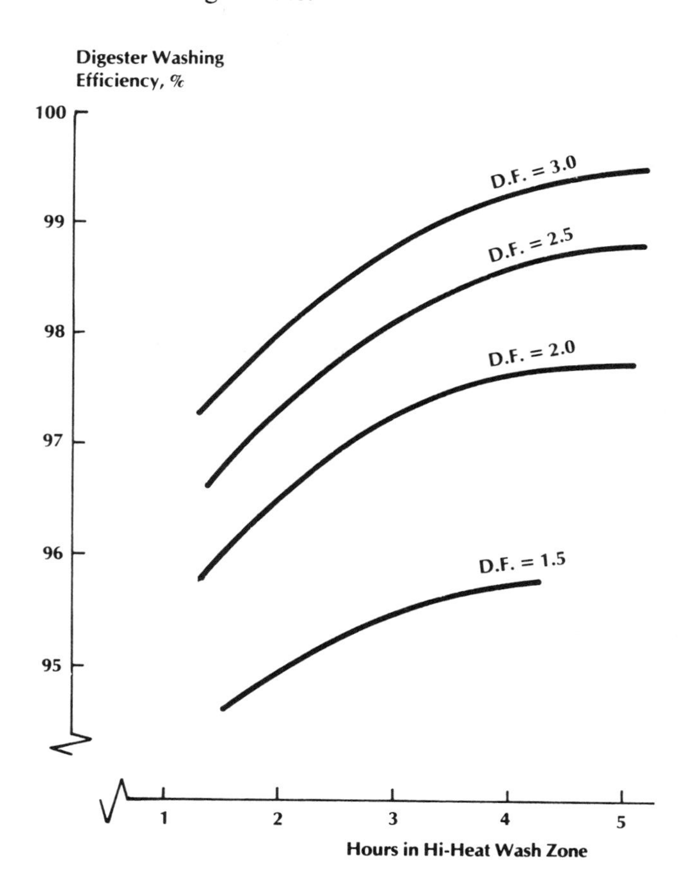

Figure 6.26 Washing efficiency in digester washing zone.

Screening

Objective of pulp screening

Pulp is screened to remove oversized and unwanted particles from good papermaking fibers so that the screened pulp is more suitable for the paper or board product in which it will be used.

The biggest oversize particles in pulp are knots. Knots can be defined as uncooked wood particles retained on a screen with 9 mm holes. The knots are removed before washing and fine screening. In high-yield pulps they are broken down in refiners and/ or fiberizers. In low-yield pulps they are removed in special coarse screens called knotters.

The main purpose of fine screening is to remove shives. Shives are small fiber bundles that have not been separated by pulping or mechanical action. They have a length of 1 mm to 3 mm, a width of 0.10 mm to 0.15 mm, and are stiffer than fibers.

Chop is another kind of oversize wood particle removed in screening. It is more of a problem when pulping hardwoods, since it originates mostly from irregularly shaped hardwood vessels and cells. Chop particles are shorter and more rigid than shives.

Debris is the name for shives, chop, and any other material that would have any sort of bad effect on the papermaking process or on the properties of the paper produced. For screening purposes, the definition of debris is any material retained on a laboratory flat screen with 0.15 mm wide slots.

Operating principles of a pulp screen

Pulp screening is an operation based on *probability*. Very few particles are so large that they cannot pass through the screen openings. Most of the particles that need to be removed can pass through the openings, but they have a lower probability of doing so than the good fibers.

The most important properties affecting the probability for passing through the screen openings are flexibility and particle length.

Shives are much stiffer and usually longer than good fibers, causing them to be rejected to a higher extent.

Figure 6.27 shows a cross section of a cylindrical screen where the pulp enters on the outside and accepts pass through the screen openings to the inside of the screen cylinder.

The movement of a particle can be divided into three kinds of movement. First, it moves inward, toward the screen. This is the *radial* velocity. Next, it moves in a circle around the screen, parallel to the screen surface. This is the *tangential* velocity. Last, it moves along the axis of the screen, from the inlet end toward the reject end of the screen plate. This is the axial velocity.

The velocity in the plane of the screen is the combination of the axial and tangential velocities. This is called the *transverse* velocity. Figures 6.27 & 6.28 illustrate the different velocities.

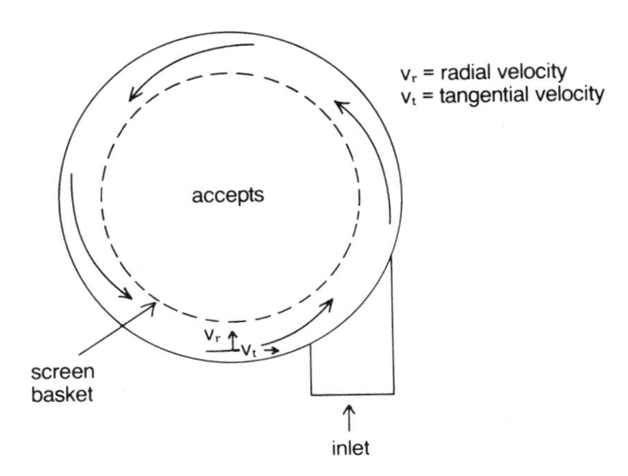

Figure 6.27 Radial and tangential movement in a screen.

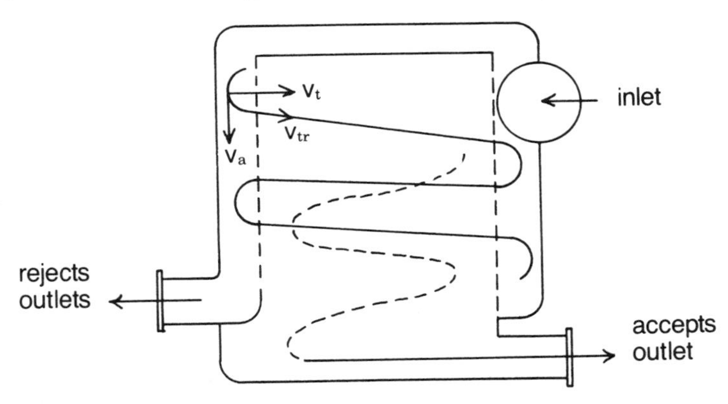

Figure 6.28 Axial and tangential movement in a screen.

The fibers and shives are mostly oriented parallel to the direction of flow and move with about the same velocity as the surrounding liquid. In order to escape through a screen opening, the flowing stock must change its direction and turn off from its fairly straight path.

The flexible fibers can bend in any direction and can easily follow the water flow through the screen openings. Stiff shives, however, cannot bend much, and they pass by the opening. If the front end of a stiff particle starts to enter the opening, its back end will "stand up." In this case, the chances are that the particle will be swept away by other stock flowing by. The stiffer and longer the particle, the greater chance it will be swept away. This is illustrated in Figure 6.29.

If the radial velocity is increased, the stock will not have to change direction as much in order to escape through the screen openings. This means that stiff particles have a greater chance of passing through. The ratio of the

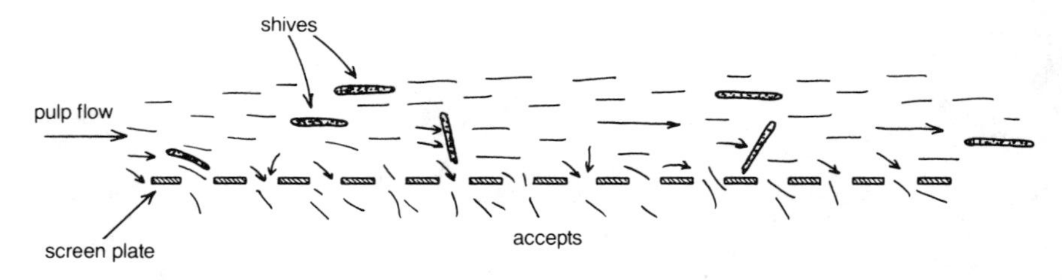

Figure 6.29 Pulp flow through screen openings.

transverse velocity to the radial velocity (V_T/V_R) is a very important parameter in determining the efficiency of separation in a screen. The higher the ratio, the more efficient the separation. Control parameters, discussed later, determine this ratio.

Different types of screens

There are three major types of stock screens that have been used throughout the years:

1. Vibratory screens
2. Centrifugal gravity screens
3. Pressure screens.

At one time, *vibratory flat screens* were the only type screen available, but they are no longer used for stock screening. Compared to screen designs available these days, they had many disadvantages: their open design gave problems with foaming; they required high maintenance; and they had a low capacity per unit area. Now, flat vibratory screens are only used as tailings screens.

The centrifugal gravity screen came into use in the fifties and is still in use today. (Figure 6.30 shows a centrifugal gravity screen.) It consists of a horizontal, cylindrical, perforated screen plate. The feed enters at one end of the cylinder. As the pulp slurry flows across the screen surface, the accepts pass through the screen openings while the rejects do not pass through. The rejects flow then exits at the opposite end of the cylinder.

Dilution water is injected through a central shaft. Dilution is necessary since the stock inside the cylinder is thickened as it flows towards the rejects outlet. A bladed rotor keeps the screen plate clean by scraping off deposited fibers. This kind of screen discharges to atmospheric pressure, meaning that foam generation is still a problem.

The pressure screen is the most recent development in stock screening and is widely used today. The operating principle is similar to that of centrifugal gravity screens. The main difference is that pressure screens are fully enclosed and designed to discharge against a positive back pressure.

There are a number of designs available, all utilizing a cylindrical perforated plate. The designs differ in screen cylinder orientation, direction of flows and type of rotor, which keeps the screen clean.

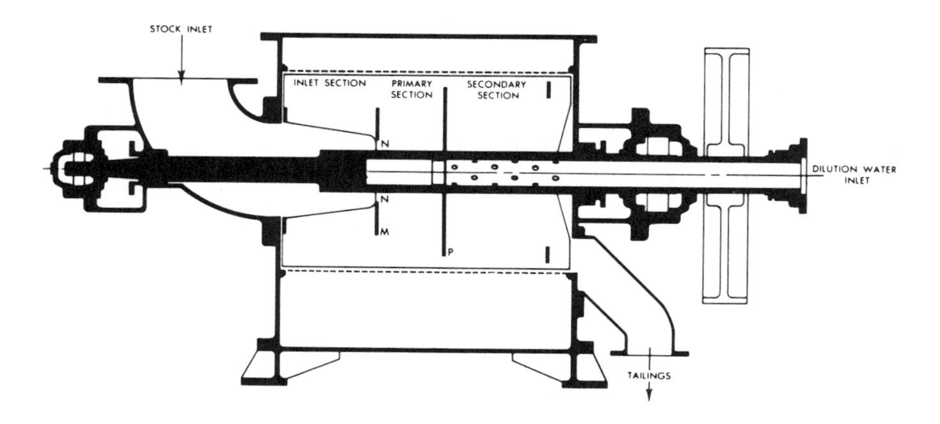

Figure 6.30 Centrifugal gravity screen.

The screen cylinder can have a horizontal orientation, like gravity screens, or can be vertically oriented, as shown in Figure 6.31.

Screen designs are available with either outward radial flow or inward radial flow. A screen is said to have an outward radial flow when the feed enters on the inside of the cylinder and the accepts flow outward, toward the periphery and through the screen openings.

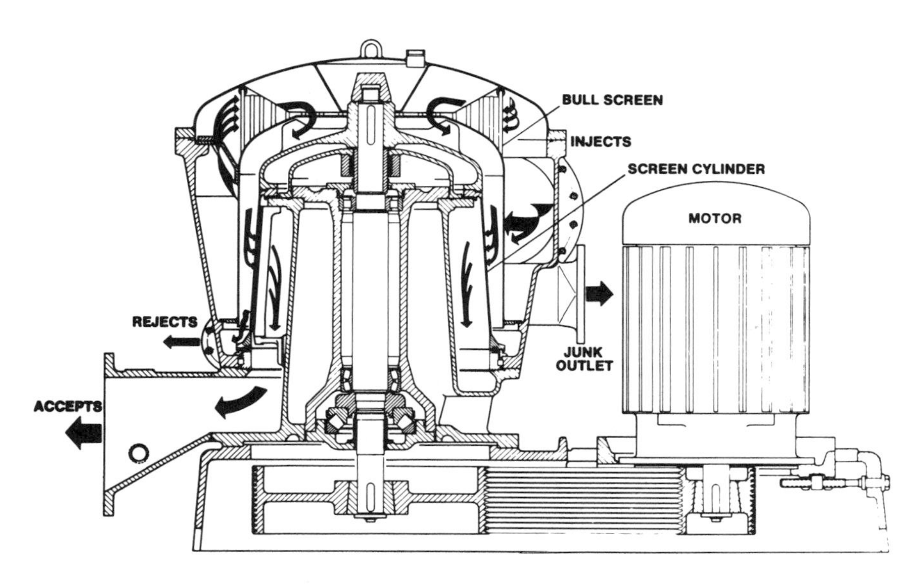

Figure 6.31 Vertically orientated pressure screen.

There are also designs in which the feed enters between two concentric screens and the accepts flow both inward and outward. Figure 6.32 shows different screen configurations.

The most common cleaning mechanism for pressure screens is rotating hydrofoils. As the foil rotates close to the screen, the fibers in front of it are pushed through the screen openings. A pressure wake is created between the screen and the back end of the foil. Fibers from the outside of the screen are, therefore, sucked back through the screen openings. This negative pulse effectively cleans the openings of deposited fibers. Figure 6.33 shows the operating principle of a hydrofoil.

Other kinds of rotors used are bladed rotors and cylindrical rotors with bumps, which create the necessary pulse across the screen.

Pressure screens can operate with or without dilution water. Pressure screens have a number of advantages compared to gravity screens:

- They have a higher capacity per area unit.
- Since they are totally enclosed, there are no problems with foaming or air entrainment.
- They require less space and have higher flexibility as to where they can be placed.
- They are not as sensitive to throughput rate, feed consistency, or reject rate as gravity screens.
- They can work at higher consistencies.

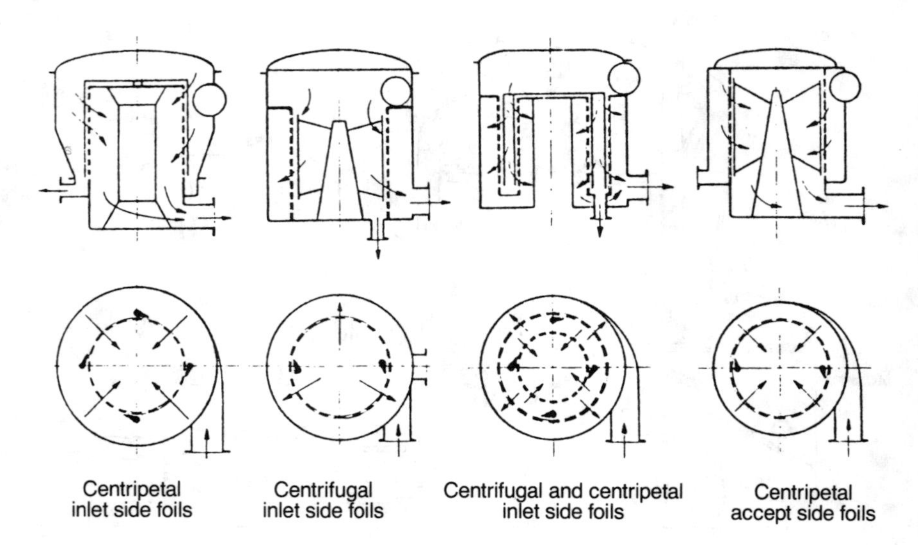

Figure 6.32 Different flow configurations.

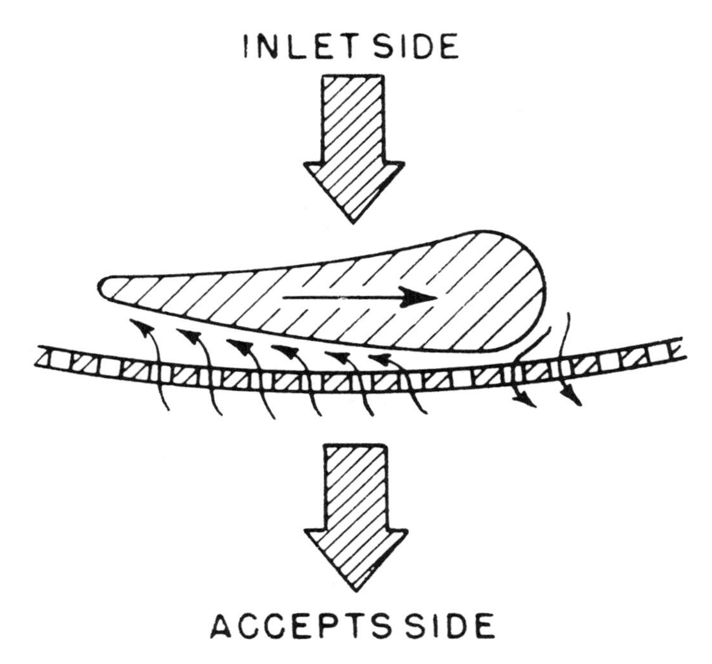

Figure 6.33 Hydrofoil.

Control parameters in pulp screening

The variables that can be used for controlling the pulp screening operation are:

- Screen design
- Reject rate
- Feed consistency
- Internal dilution
- Rotor speed.

The screen plate has a larger influence on operation than any other controllable parameter but, obviously, it cannot be varied during operation.

Reject rate, feed consistency, and internal dilution are the three parameters that the screen operator will use for continuous control of the screening operation. The rotor speed can also be varied but is usually kept constant during operation.

Screen plate design

Screen type and size of screen openings are very critical control parameters, although they cannot be varied during operation of the screen. It is possible to change screens in a short time, however, and there are mills that change plates when they change grades or fiber furnishes.

The screen plate parameters are:

- Screen plate size (size of screen openings)
- Percent open area
- Screen plate type (grooved or smooth)
- Type of screen plate openings (round or slotted).

Screen plate size affects the capacity, efficiency, and minimum reject rate at which the screen will operate. The capacity increases with increasing screen plate size, while efficiency and minimum reject rate decrease. Screen plate size is also dependent on the type pulp furnish that is being screened. Hardwoods require smaller screen openings, since hardwood fibers are smaller than softwood fibers.

Percent open area depends on hole diameter and the distance between adjacent holes. As hole size is decreased, the distance between the holes must also decrease in order to keep the open area identical. If the holes are too close, the fibers can enter two holes at once, causing fiber build-up, called "stapling." Therefore, when using smaller holes for screening softwood pulp, a lower open area must be used. The normal range for open area is 12% to 23%.

There are two screen plate types, smooth screens and grooved screens. Grooved screens have come into use in recent years and are the most common nowadays. (See Figure 6.34.)

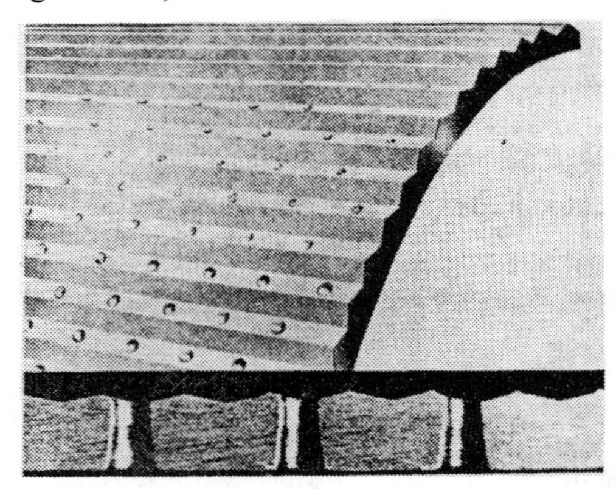

Figure 6.34 Grooved screen surface.

The grooves will increase turbulence, which increases capacity. For softwood, a grooved plate has 25% to 50% higher capacity than a smooth plate of the same size. Smaller holes can, therefore, be used with grooved plates without loss of capacity.

At the same reject rate, grooved plates have a slightly lower efficiency than smooth plates of the same size. They tend to reject less material and control settings must, therefore, be altered to maintain the reject rate if changing from smooth to grooved plates.

Softwood is more affected by grooved plates than hardwood, because grooving has a greater effect on long fibers. The increase in capacity is an effect of more long fibers passing through the screen and becoming accepts.

Round holes and narrow slots are the two main kinds of screen openings being used.

Slotted screens are becoming increasingly popular for screening kraft pulps. This is because they have a higher shive and speck removal efficiency than screens with round holes. Figure 6.35 shows a section of a slotted screen.

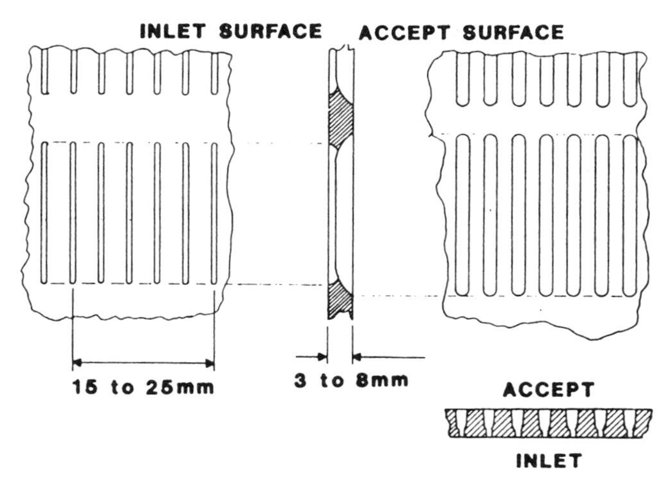

Figure 6.35 Section of slotted screen.

Slotted screens have a lower capacity than screens with holes of normal size, however. Manufacturers are trying to overcome this by combining slots with grooves or other kinds of profile designs.

Reject Rate

Reject rate is the major control parameter the screen operator uses for continuous control of the screening operation. By regulating control valves, the operator can directly control the *volume* reject rate. The *mass* reject rate change is not proportional to the volume reject rate change, since the accept flow and reject flow consistencies will change as well. For example, when reducing the rejects flow, the rejects consistency will increase, counteracting the intended decrease in mass reject rate.

Once the relationship is known between volume reject flow and mass reject rate for the specific conditions used, the mass reject rate can be controlled fairly well within limits. By increasing reject rate, the screen efficiency is increased. This is illustrated in Figure 6.36. The accepts flow will contain less shives and other debris. Reject rates cannot be set too high, however, since more good fiber is rejected at higher reject rates.

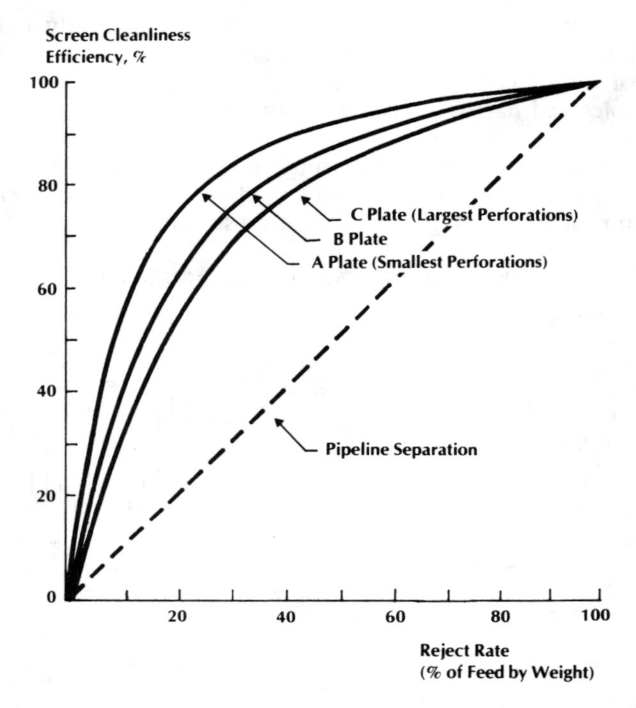

Figure 6.36 Effect of reject rate on screening efficiency.

Every screen has a minimum reject rate for a given set of pulp conditions. At reject rates above the minimum rate, the screen will operate continuously and efficiently. The screen can operate at reject rates lower than the minimum rate, but the efficiency will be low and the risk for plugging much higher. The minimum reject rate is not a constant. It depends on many variables: feed temperature, feed debris, and feed consistency, for example.

Feed consistency

Centrifugal gravity screens are sensitive to feed consistency. The efficiency will be poor if feed consistency is too different from design consistency. Pressure screens are not as sensitive. Each pressure screen has a range of feed consistencies over which it will operate at normal efficiency, using the same reject rate.

For a given fiber throughput, a lower feed consistency means a higher volume flow. A higher volume flow means higher radial velocities of the fibers and a higher pressure drop across the screen. Each screen has a maximum operational pressure drop, and screen capacity is limited by decreasing feed consistency.

If the radial velocity is too high, the stiff shives will approach the screen openings at an angle that will let them pass through more easily. The efficiency is, therefore, decreased if the volume feed rate is too high. If the screen is operated at consistencies above the design range, the reject rate must be raised in order to maintain runability and efficiency. In earlier days, screens operated at consistencies of one to two percent. Modern pressure screens operate at higher consistencies, in the range of 2% to 5%.

Internal dilution

In centrifugal gravity screens, the internal dilution controls the mass reject rate. It washes good fiber into the accepts, decreasing the rejects flow across the screen plate. Pressure screens do not always have internal dilution. Internal dilution is used only to control the consistency of the stock inside the screen and the rejects consistency. Excessive internal dilution in a pressure screen will always decrease capacity and can seriously decrease efficiency.

Evaluation of screen performance

The performance of a stock screen is usually expressed in terms of its ability to remove debris consisting mostly of shives.

The screening efficiency (E) is defined as:

$$E = \frac{S_i - S_a}{S_i}$$

S_i = percent debris by weight in feed
S_a = percent debris by weight in accepts

Percent debris is defined as material retained on a laboratory flat screen with 0.15 mm slots. It is important to note that this equation is not based on a mass balance over the screen, since the mass flow rates of accepts and feed is not part of the equation. The screening efficiency is very dependent on how much stock is rejected, as can be seen in Figure 6.36. Therefore, when comparing the screening efficiency of two different screens, their reject rates must also be considered.

Pulp screening systems

In order to reach an acceptably low level of shives in the accepts stream, the reject rate must be fairly large, about 15% to 20%. A number of good fibers will follow the rejects stream out, and they must be recovered and returned to the process. This is achieved by screening in multiple stages, where up to four stages of screening is used. The most common arrangement of screens is the so-called cascade screening system. A three-stage cascade screening system is illustrated in Figure 6.37.

The rejects from the primary screens are fed to the secondary screens. The accepts from the secondary screen are recirculated to the feed of the primary screen, while the rejects are sent to the tertiary screen. Its accepts stream is

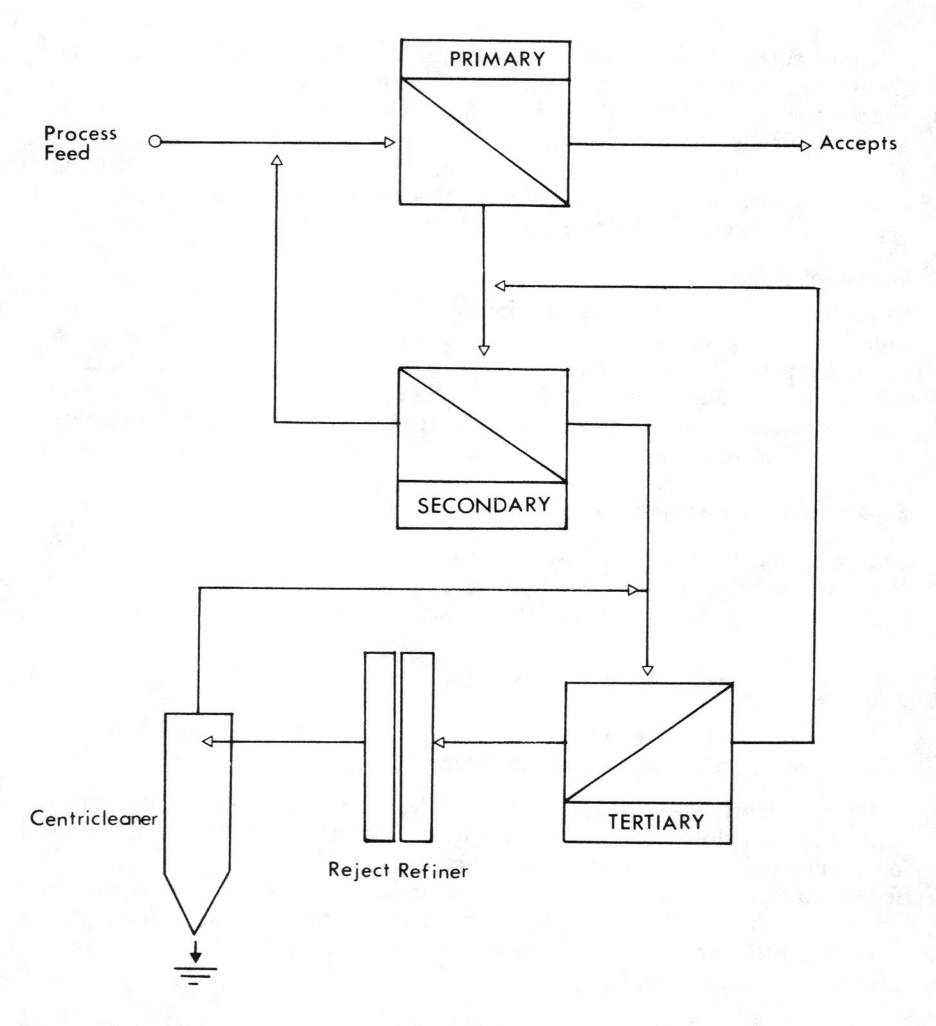

Figure 6.37 Three-stage cascade screening system.

fed to the secondary screen, while the final rejects are treated, depending on what grade is being produced.

If a very high pulp quality is required, the rejects are thickened and removed from the system. They can then be sent to a landfill, burned in a hog fuel boiler, recooked, or sent to a high-yield line, if the mill has one.

If quality requirements are not extremely high, the rejects can be refined in a rejects refiner. The refined rejects are often cleaned in a centrifugal cleaner in order to prevent a buildup of trash and grit. The refined rejects are usually returned to the feed of the last stage screen.

Sometimes the quality required cannot be achieved with one stage of primary screening. A second primary screen is then used for rescreening the accepts from the first primary screen. The rejects from the second primary screen are usually sent to the secondary screen.

Centrifugal Cleaning

Objective of centrifugal cleaning

The objective of centrifugal cleaning is to remove small unwanted particles still remaining in the pulp after screening. This undesirable material consists of sand and grit, bark, small shives, fly ash, rust scales, and all other constituents, which would lower the appearance of the final paper product if not removed.

It is normally only high-quality bleachable pulp stocks that are cleaned. For lower grades, such as linerboard, where appearance is not so important, centrifugal cleaning is usually not applied.

Operation principles

There are many names for the basic unit used in centrifugal cleaning: centricleaner, centrifugal cleaner, cyclone, or hydrocyclone.

The unit is a tall, thin cylindrical vessel with a conical bottom. (See Figure 6.38.)

The pulp stock enters the centrifugal cleaner through a tangential inlet near the top. Its tangential motion is converted to a rotating motion and the stock starts to flow downward in a spiral.

The outlet for the accepts part of the pulp stock is located at the top, in the center of the unit. Thus, in order to escape through the accepts inlet, the rotating stock is forced to spiral inward, toward the center. Consequently, each particle in the stock has two opposing forces acting on it: the centrifugal force, which wants it to move out towards the periphery, and the drag force created by the inward flowing stock, which escapes through the accepts outlet.

As a particle moves inward with the accepts stock flow, the velocity increases, resulting in a higher centrifugal force. Denser particles cannot overcome this centrifugal force and will not be able to escape through the accepts outlet. Instead, they are carried away with the downward spiral flow. Closer to the bottom, the diameter narrows, and the flow is forced inward, again increasing speed and centrifugal force. This concentrates the dirt and releases good fibers to the accepts flow. The rejects flow leaves the centrifugal cleaner through an outlet located at the bottom.

The flow pattern in the described centrifugal cleaner is "a spiral within a spiral." At the periphery, there is a downward spiral flow ending up at the rejects outlet and, closer to the center, is an upwards spiral flow that exits through the accepts outlet. In the center of the cleaner is an air core. This flow pattern is illustrated in Figure 6.39.

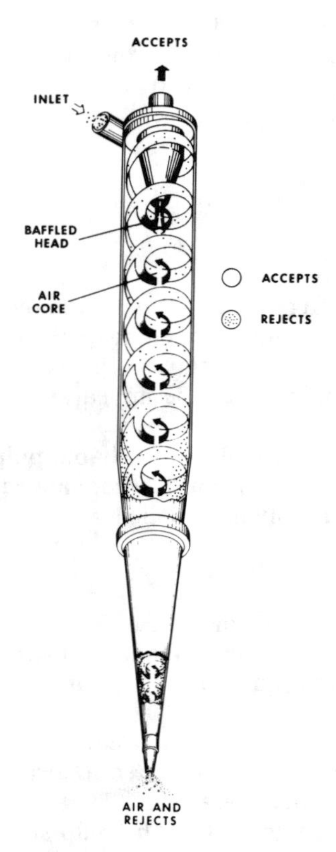

Figure 6.38 Schematic of a centrifugal cleaner.

The centrifugal cleaner removes unwanted particles by a combination of centrifugal force and fluid shear. Therefore, it separates not only on the basis of differences in density but also, to some extent, particle shape.

There is another type of centrifugal cleaner called "the reverse hydrocyclone." This kind of cleaner has the same basic design and operates based on the same principles as conventional centrifugal cleaners. In these cyclones, the centrifugal forces acting on undesirable particles like plastics are, on the average, *smaller* than the corresponding drag forces. Consequently, these particles will, together with some fiber, travel inward and upwards and escape through the top outlet. This fraction is now the rejects fraction, while the accepts fraction of fibers only leaves through the bottom outlet.

Since the rejects flow will consist of particles less dense than the accepts flow, the reverse hydrocyclone is efficient in removing flaky plastic fragments, hot smelts, glues, waxes, and styrofoam. These contaminants are all very common in secondary fiber processing, and most reverse cyclone systems are installed in secondary fiber mills.

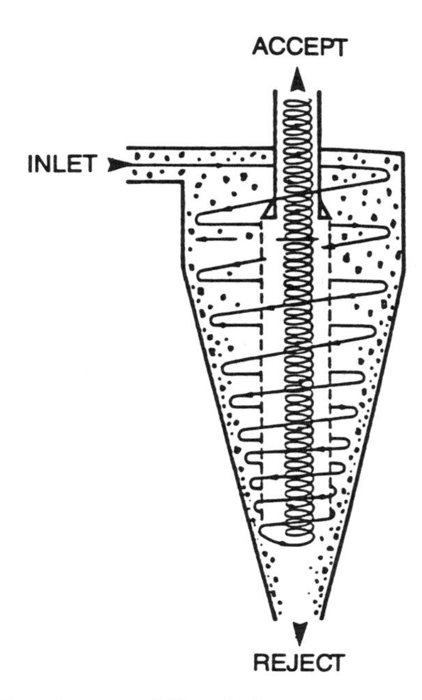

Figure 6.39 Flow pattern in a centrifugal cleaner.

Centrifugal cleaner systems

Centrifugal cleaners are often installed in a cascade sequence similar to that used for pulp screens. (See Figure 6.40.) This is because the rejects flow for individual cleaners must be kept at a level of 10% to 30% in order to obtain a good operating efficiency. This reject flow contains good fibers that must be recovered in subsequent stages. Accepts from the following stages are usually recirculated to the feed of the preceding stage, creating a true cascade arrangement.

As for pulp screens, other arrangements are frequently used. Two to four stages of screening are normal.

A typical centrifugal cleaner has a base diameter ranging from 5 cm to 30 cm and a nominal feed flow of 40 L to 500 L per minute. Many units, therefore, must operate in parallel within each cleaning stage. The individual centrifugal cleaners are often grouped in housings of different designs, saving space and making installation easier. (See Figure 6.41.)

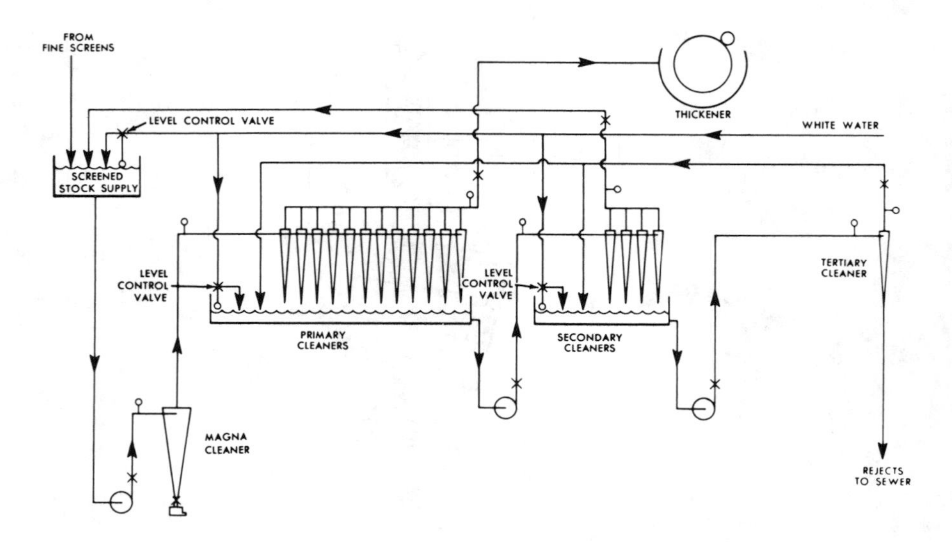

Figure 6.40 Flowchart for a cleaner system.

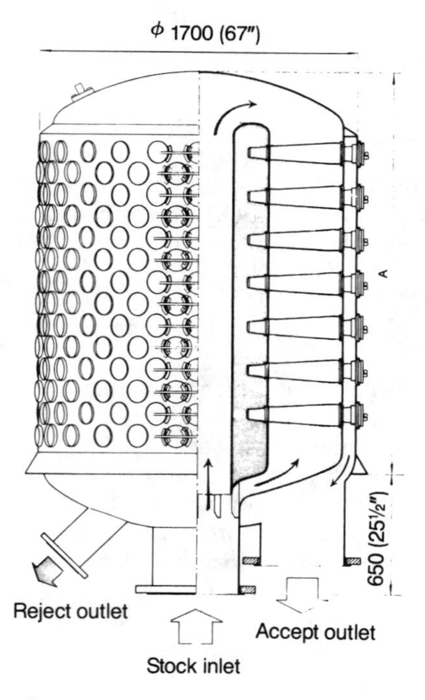

Figure 6.41 Housing for cleaners.

Operating variables

The performance of a centrifugal cleaner is usually measured in terms of its ability to remove dirt particles.

The definition of cleaning efficiency then is:

$$\text{Cleaning efficiency }\% = \frac{(\text{dirt count in feed} - \text{dirt count in accepts})}{\text{dirt count in feed}} \times 100$$

The most important operating variables are feed consistency, flow rate, pressure drop across the unit, and reject rate.

Centrifugal cleaning is usually carried out at low consistencies, between 0.25% and 1.5%. Optimum efficiency is normally reached between 0.35% and 0.5% and from there, it drops considerably with increasing consistency. This is illustrated in Figure 6.42.

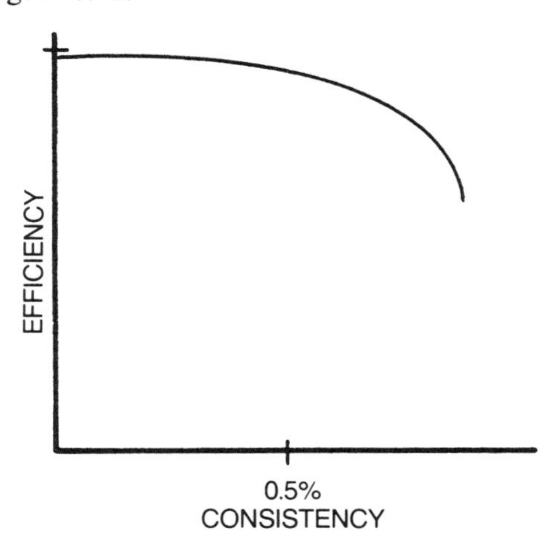

Figure 6.42 Cleaning efficiency vs. consistency.

The pressure drop determines the hydraulic capacity of the cleaner and is a measure of how much centrifugal action is developed. Most units operate with a pressure drop of 30 psi to 35 psi. As can be seen in Figure 6.43, pressure drop has a big impact on efficiency as well.

Pressure drop or feed flow can easily be adjusted on all types of centrifugal cleaners. On the other hand, for many cleaners, reject rate is a function of the reject outlet diameter. Therefore, adjustments of reject rate can only be made by replacing the reject outlet nozzle. This design of centrifugal cleaners is often referred to as "conventional" cleaners, and they normally discharge rejects against atmosphere pressure.

In newer, completely closed cleaner systems with many units arranged in housings, the reject rate can be varied. In these systems, the rejects are

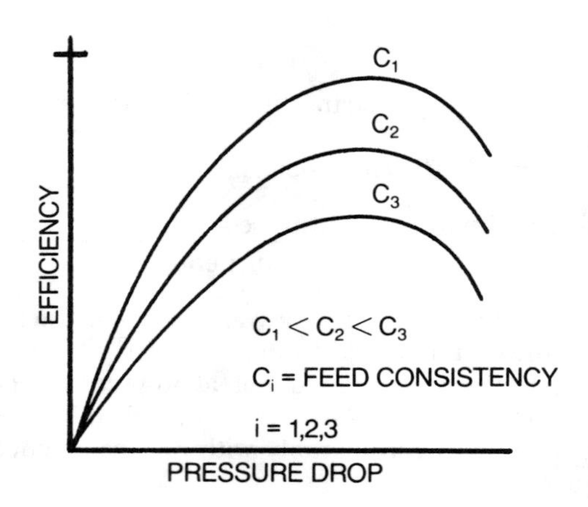

Figure 6.43 Cleaning efficiency vs. pressure drop.

discharged to a common, completely filled and pressurized reject chamber, making reject flow adjustments possible simply by throttling a valve in the reject pipe.

Operating problems

The main operating problem with centrifugal cleaners is that the rejects outlet may become plugged with foreign materials, fiber flocks, or high-consistency stock. This will reduce the system efficiency and cause very rapid wear of the plugged cleaner.

The rejects outlet is usually sized to maintain a reasonable reject level, but sometimes it must be made bigger in order to avoid plugging. Lowering the feed consistency is another way to avoid plugging. Systems operating with pressurized rejects chambers are less susceptible to plugging. Because of the pressurization, the reject outlets can be larger than the feed holes, virtually eliminating plugging. Another feature lowering the risk of plugging is "action water." Dilution water is tangentially introduced into the reject nozzles of individual cleaners, boosting the centrifugal action and diluting the reject fraction.

Thickening

Objective of thickening

After screening and cleaning, the pulp has a low consistency. Centrifugal cleaning, for example, is carried out at about 0.5% consistency, which corresponds to one ton of pulp per 199 tons of water. The pulp slurry must be thickened by removing some of the water before further processing. The reasons for thickening are:

- *The cost of storage*: Low-consistency pulp requires bigger storage tanks than higher-consistency pulps for storing the same amount of pulp fiber.
- *Easier control of consistency required for further processing*: It is difficult to control the consistency of very dilute suspensions of pulp.

Pulp storage is generally termed low- or high-density. Low density corresponds to a pulp consistency range of 2.5% to 4%. At these consistencies, the pulp can be pumped and agitated. High density corresponds to a consistency in the 10% to 14% range. This pulp cannot be blended by agitation, but it can still be pumped by displacement pumps or special centrifugal pumps.

There is a variety of equipment for thickening, depending on what level of consistency is required. (See Figure 6.44.)

Equipment	*Discharge Consistency*	*Washing Capability*
Slusher	3.5 - 4.0%	none
Gravity Thickener	4 - 8%	none
Valveless Filter (Internal Dropleg)	9 - 12%	some
Vacuum Filter	12 - 15%	large
Multi-disc Filter	10 - 12%	none
Screw Extractor	>20%	none
Various Press Designs	>20%	none → large

Figure 6.44 Equipment for thickening.

Low-density thickening

The *decker* or *gravity thickener* is the most common type of thickener. It consists of a wire mesh covered cylinder rotating in a vat and resembles a brown stock drum washer in operation. (See Figure 6.45.)

In the decker or gravity thickener, the feed is supplied to the vat uniformly along the length of the cylinder. Water drains through the wire mesh into the interior of the cylinder and is removed. The driving force is the difference in liquid level between the vat level and the water level inside the cylinder. The fiber remaining on the surface forms a sheet that is transferred from the

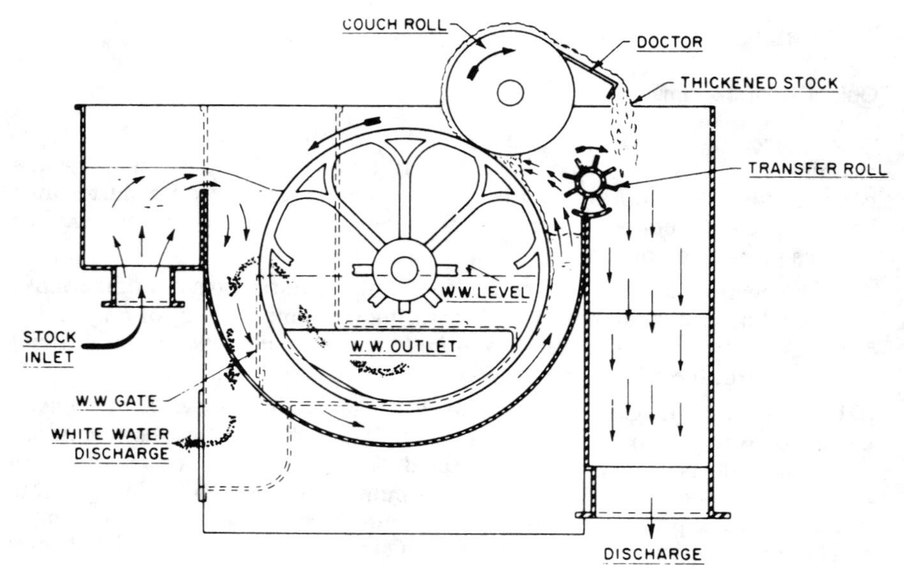

Figure 6.45 Decker.

cylinder to a couch roll. It is scraped off the couch roll by a doctor blade, and the pulp is discharged at a consistency of 4% to 8%.

The operating variables are the drum speed and the difference in liquid level that accounts for the driving force. When the drum speed is too low, the fiber mat becomes so thick that further drainage and mat build-up is prevented while still in the vat. When drum speeds are too high, fiber will wash off the wire, resulting in a bad filtering effect of the white water and low capacity. Optimum drum speed occurs when the wire emerges from the stock as soon as the optimum thickness of mat has been formed.

A slusher is a thickener that can increase consistency to about 4%. It is similar to the decker, but has no couch roll. The stock moves from the inlet side, through the dewatering zone, to the other side of the vat, where the thickened stock is discharged, often with the help of a shower. (See Figure 6.46.)

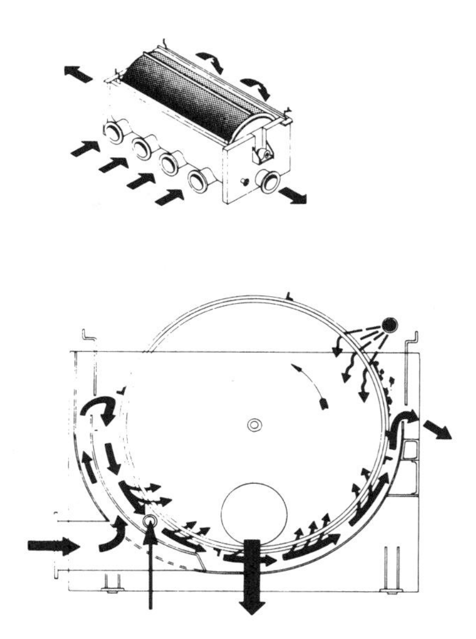

Figure 6.46 Slusher.

High-density thickening

To achieve consistencies above 10%, some kind of vacuum thickener can be used (see Figure 6.47). Most designs resemble the decker with the difference being that vacuum is applied to increase the drainage rate and maximum obtainable consistency. Some vacuum thickeners also include a washing stage. They are identical to a rotary vacuum washer. The vacuum may be developed by a vacuum pump or a barometric dropleg. Because of the larger pressure differential across the wire, the drainage capacity per unit area of wire is much larger for vacuum thickeners than for the gravity decker. Consistencies up to 15% can be achieved with these designs.

To achieve consistency levels above 15%, some type of screw thickeners or presses must be used. In an *inclined screw thickener,* an inclined screw moves the pulp upwards in a perforated inclined cylinder, the water drains through the perforation, and the thickened pulp is discharged at the top. (See Figure 6.48.)

The *twin-roll press* consists of two horizontal porous rolls, mounted in a sealed vat, that are rotated at the same speed toward each other. Pulp at a consistency of 2% to 5% is pumped into the vat. Water drains through the roll surface, forming a mat of pulp, which is carried into the nip by the rotation of the rolls and pressed to the desired consistency. The thickened stock is doctored from the rolls and discharged at high consistency. (See Figure 6.49.)

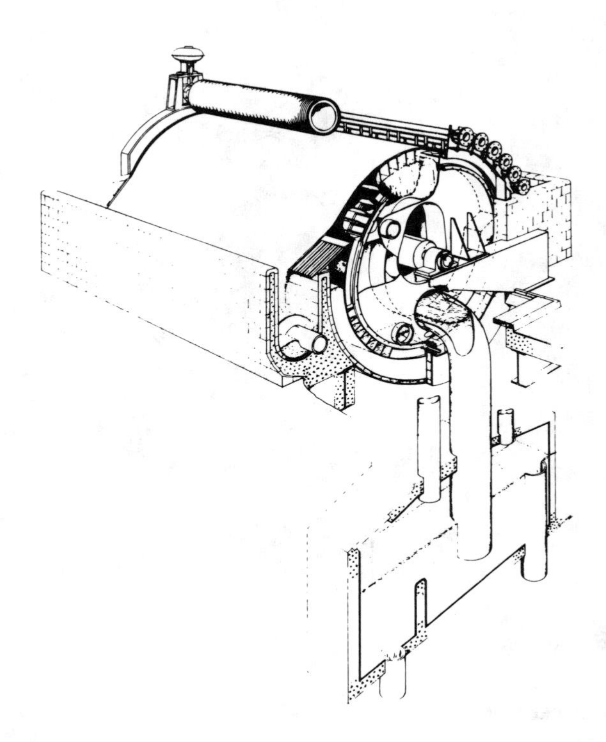

Figure 6.47 Vacuum thickener including washing.

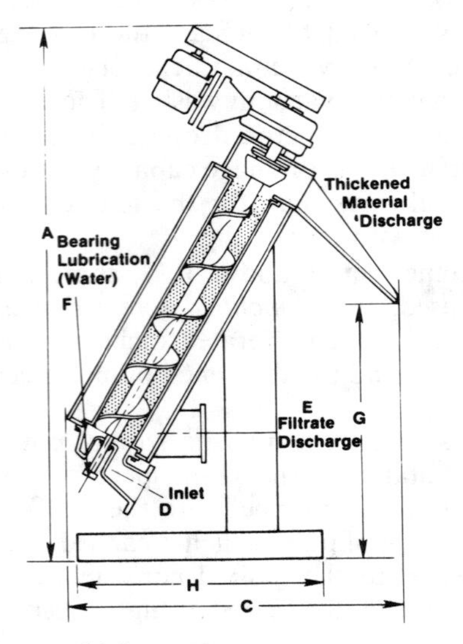

Figure 6.48 Inclined screw thickener.

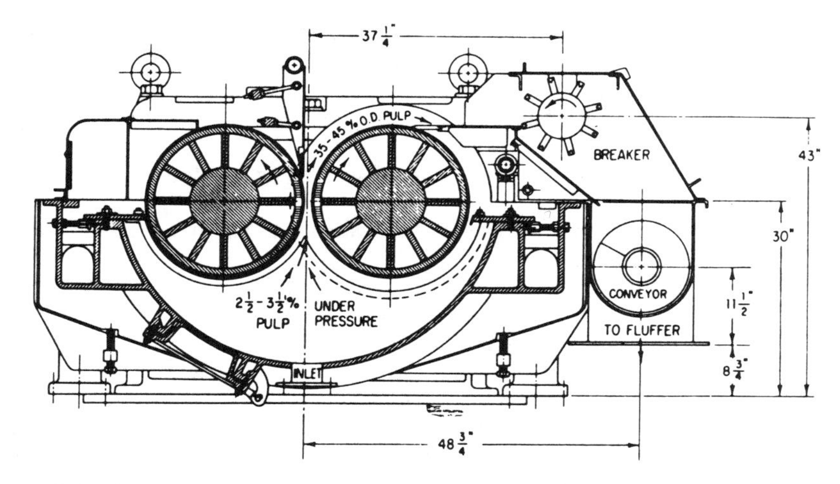

Figure 6.49 Twin-roll press.

When thickening from a very low consistency to a level of 10% to 12%, a two-stage system is often used, for example, a slusher followed by a vacuum thickener.

Questions

1. Distinguish between a high-yield pulp processing system and a low-yield pulp processing system.
2. Describe the layout and operation of two different kinds of linerboard pulping systems.
3. Discuss the treatment of knots and screen rejects in medium-yield pulp processing systems.
4. Discuss the location of washers and screens in a low-yield pulp processing system.
5. Describe the design and operation of refiners and prebreakers.
6. State the objective of brown stock washing.
7. Explain the difference between dilution/extraction washing and displacement washing.
8. Define the following terms: soda loss, dilution factor, and displacement ratio.
9. Name and identify at least four different types of washing equipment.
10. Describe the design and operation of a rotary vacuum washer.
11. Name some operating variables for rotary washers.
12. Describe the design and operation of at least one type of diffusion washer.
13. Describe the design and operation of a horizontal belt washer.
14. State the objective of pulp screening.

15. Explain why pulp screening is an operation based on probabilities.
16. Discuss the operating principles of a pulp screen.
17. Describe the design and operation of a pressure screen.
18. List the main differences in design between a gravity screen and a pressure screen.
19. Name a few reasons why vibratory screens are no longer in common use for pulp screening.
20. Name at least three variables that can be used for controlling the pulp screening operation.
21. Write the equation for screening efficiency.
22. Describe the operation of a cascade screening system.
23. State the objective of centrifugal cleaning.
24. Discuss the operating principles for a centrifugal cleaner.
25. Describe the design of a typical centrifugal cleaning system.
26. Identify the most important operating variables for a centrifugal cleaner.
27. Discuss the most common operating problem of centrifugal cleaning.
28. State the reasons for pulp thickening.
29. Define the terms "low-density storage" and "high-density storage."
30. Describe the design and operation of a decker.
31. Name two types of equipment that can be used for thickening to a pulp consistency above ten percent.

Principal Sources For Figures

Crotogino, R.H., Poirier, N.A., and Trinh, D.T. "The Principles of Pulp Washing." *Tappi J.* 70(6): (June 1987):95.

Hamilton, F., and Leopold, B., eds. *Pulp and Paper Manufacture.* Vol. 3: *Secondary Fibers and Non-Wood Pulping.* 3rd ed. Atlanta and Montreal: Joint Textbook of the Paper Industry, 1987.

Ingruber, O.V., Kocurek, M.J., and Wong, A., eds. *Pulp and Paper Manufacture.* Vol. 4: *Sulfite Science and Technology.* 3rd ed. Atlanta and Montreal: Joint Textbook of the Paper Industry, 1985.

Leask, R.A., and Kocurek, M.J., eds. *Pulp and Paper Manufacture.* Vol. 2: *Mechanical Pulping.* 3rd ed. Atlanta and Montreal: Joint Textbook of the Paper Industry, 1987.

Perkins, J.K., ed. *Brown Stock Washing Using Rotary Filters.* Atlanta: TAPPI PRESS, 1983.

Smook, G.A. *Handbook for Pulp and Paper Technologists,* Edited by M.J. Kocurek. Atlanta and Montreal: Joint Textbook Committee of the Paper Industry, 1982.

TAPPI, *Introduction to Pulping Technology,* TAPPI Home Study Course, no. 2. Atlanta: TAPPI PRESS, 1976.

TAPPI 1983 Pulping Conference. Houston, Texas: October 24–26.

TAPPI 1984 Pulping Conference. San Francisco, California: November 12–14.

TAPPI 1986 Pulping Conference. Toronto, Ontario: October 26–30.

Fig. 6.2 Kamyr Brochure.
Fig. 6.3 TAPPI, *Introduction to Pulping Technology,* Fig. 14.10, p. XIV-6.
Fig. 6.5 TAPPI 1983 Pulping Conference, Fig. VI, p. 182.
Fig. 6.6 Jones Div., Beloit Corp. Brochure.
Fig. 6.7 Black Clawson Brochure.
Fig. 6.8 Black Clawson Brochure.
Fig. 6.9 Crotogino, Poirier, and Trinh, Fig. 1, p. 96.
Fig. 6.10 Crotogino, Poirier, and Trinh, Fig. 2, p. 96.
Fig. 6.12 Smook, Fig. 9.21, p. 99.
Fig. 6.13 Perkins, Fig. 2.1, p. 7.
Fig. 6.14 Perkins, Fig. 3.5, p. 14.
Fig. 6.15 Smook, Fig. 9.5, p. 92.
Fig. 6.16 Smook, Fig. 9.15, p. 97.
Fig. 6.17 Smook, Fig. 9.16, p. 97.
Fig. 6.18 Kamyr Brochure.
Fig. 6.19 Smook, Fig. 9.12, p. 96.
Fig. 6.20 Smook, Fig. 9.13, p. 96.
Fig. 6.21 Kamyr Brochure.
Fig. 6.22 Smook, Fig. 9.17, p. 98.
Fig. 6.23 Smook, Fig. 9.20, p. 98.
Fig. 6.24 Leask, Fig. 179A, p. 224.
Fig. 6.25 Smook, Fig. 9.19, p. 98.
Fig. 6.26 Smook, Fig. 9.9, p. 95.
Fig. 6.30 Smook, Fig. 9.22, p. 100.
Fig. 6.31 Smook, Fig. 9.23, p. 100.
Fig. 6.32 Smook, Fig. 9.25, p. 102.
Fig. 6.33 Smook, Fig. 9.26, p. 103.
Fig. 6.34 TAPPI 1984 Pulping Conference, Fig. 8, p. 518.
Fig. 6.35 TAPPI 1986 Pulping Conference, Fig. 2, p. 279.
Fig. 6.36 Smook, Fig. 9.27, p. 103.
Fig. 6.37 Smook, Fig. 9.28, p. 104.
Fig. 6.38 Smook, Fig. 9.30, p. 105.
Fig. 6.39 Leask, Fig. 154, p. 204.
Fig. 6.40 Smook, Fig. 9.33, p. 107.
Fig. 6.41 Hamilton and Leopold, Fig. 192, p. 229.
Fig. 6.44 Smook, Table 9.5, p. 109.
Fig. 6.45 Smook, Fig. 9.37, p. 108.
Fig. 6.46 Smook, Fig. 9.38, p. 108.
Fig. 6.47 Smook, Fig. 9.42, p. 110.
Fig. 6.48 Black Clawson Brochure.
Fig. 6.49 Ingruber, Kocurek, and Wong, Fig. 227, p. 251.

7 The Kraft Recovery Process

Overview

The recovery of the cooking chemicals used in the kraft pulping process has always been an economic necessity. This is in contrast with sulfite pulping where disposing of the spent liquors to streams was common until environmental regulations prohibited that. Kraft recovery systems recover the inorganic chemicals of sodium hydroxide and sodium sulfide, and burn the dissolved organic materials as fuel to produce steam and electricity. The primary operations that constitute kraft chemical recovery include evaporation, burning, and causticizing.

Evaporation concentrates the black liquor from about 14% to 18% solids to a level of about 68% to 70% or higher. Multiple-effect evaporators, either long tube vertical or falling film type, are used. Concentrators may be used to further increase solids from the evaporators. Older technology employed direct contact evaporators which utilized the hot flue gases from the furnace. Tall oil is recovered, both during and after evaporation. Scaling of the evaporator tubes is a major operating problem.

Burning occurs in a recovery furnace. About 45% to 55% of the wood entering the digester ends up as fuel. Combustion air is introduced to the furnace at three levels. Two air entries are located below the firing level for the black liquor and maintain the reducing atmosphere required to reduce the oxidized sulfur compounds present (mainly sodium sulfate and thiosulfate) to sodium sulfide. The third or upper level of combustion air completes the burning. Steam and electricity are generated from the heat produced. Leaving the furnace in the form of a molten smelt are sodium sulfide and sodium carbonate. When dissolved in water, the smelt forms green liquor.

Causticizing converts the green liquor to white liquor. Sodium carbonate, Na_2CO_3 is reacted with lime, CaO, to form sodium hydroxide, NaOH. This produces white liquor and a lime mud, $CaCO_3$. Calcining in a kiln converts the mud back to calcium oxide. Causticizing includes smelt dissolving, green liquor clarification, slaking with lime, causticizing, white liquor clarification, mud washing, and kiln operations.

Functions and steps

Kraft recovery involves these major functions:

- Burning the organic substances (mostly lignin and simple carbohydrates) dissolved during pulping. This supplies a major portion of the heat needed for most kraft mills and eliminates a major source of stream pollution.
- Regenerates the sodium hydroxide (NaOH) needed for pulping.

- Converts the sulfur compounds in the black liquor to sodium sulfide (Na_2S) which is the other required pulping chemical.

The main steps in the kraft recovery process are:

- Concentration of the "weak black liquor," or "spent liquor," obtained from the pulp washers by evaporation to "strong black liquor"
- Burning the concentrated black liquor in a furnace
- Dissolving the smelt from the furnace to form green liquor
- Causticizing the green liquor with lime to form white liquor
- Burning the lime mud to recover lime.

The kraft recovery process is cyclic in nature, as shown in Figure 7.1.

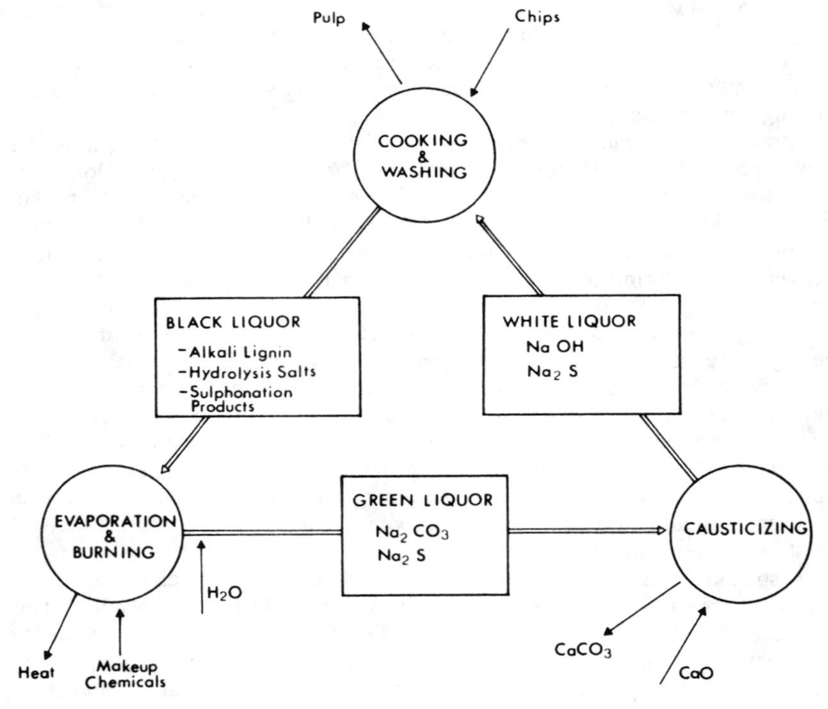

Figure 7.1 Kraft liquor cycle.

Some losses of chemicals always occur in pulping. Sodium salts can be lost due to insufficient washing of the pulp. Some sodium is bound (sorbed) to the pulp and cannot be removed even with excellent washing. Sulfur compounds are lost mainly in the gas phase; in digester relief and blow gases, in volatile gases during evaporation and in the flue gases from recovery. Nevertheless, a modern kraft mill recovers about 95% to 97% of the chemicals used.

Saltcake (sodium sulfate) is the traditional make-up chemical used. Indeed, another name for the kraft process is the sulfate process. In bleached kraft mills, the spent chlorine dioxide generator liquor, containing sulfates, is a major source of make-up. Environmental regulations forced closure of the mills, reduced chemical losses and resulted in imbalances in the ratio of sulfur to sodium. Today, several other chemicals are also commonly used: elemental sulfur, spent sulfide-containing liquors (as from petroleum refining), and soda ash (sodium carbonate).

Black Liquor Processing

Composition of a typical black liquor

Weak black liquor from pulp washing normally contains about 14% to 18% solids. There is some unconsumed caustic and the pH is typically 12-plus. About 65% of the solids in the black liquor are organic in nature, originating from the wood used; 35% are inorganic, coming from the chemicals in the cooking (white) liquor. The organic compounds are in the form of sodium salts of lignin, resin and fatty acids, acids from carbohydrates, and miscellaneous organic compounds found in the wood used. The inorganic chemicals are sodium carbonate, sodium thiosulfate, sodium sulfide, and sodium hydrosulfide. The heating value of the black liquor solids is the sum of the heating values of each of these components; typically, 5,400 to 6,500 BTU's per lb of black liquor solids.

Black liquor evaporation

Objective

The objective of black liquor evaporation is to concentrate the solids in the liquor from about 14% to 18% to about 70% or more. To accomplish this, about 80% of the water must be removed. The remaining water will be removed in the recovery furnace. To burn black liquor safely, the solids content must be at least 60%; many mills require a minimum of 62% or higher.

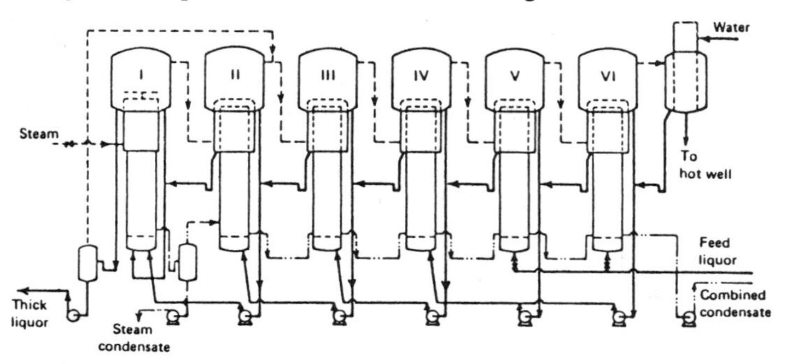

Figure 7.2 Multiple-effect evaporators

Multiple-effect evaporators

The multiple-effect evaporators are a series of bodies or units (called effects), usually 5 to 7, where water is evaporated from the black liquor (see Figure 7.2). These effects are operated at different pressures so that the vapor formed when the liquor boils in one effect can be used as heating steam in the next effect operating at a lower pressure and temperature. The effects are always numbered in the direction of decreasing steam temperature. Fresh steam enters the first effect while the lowest pressure steam exits the last effect. The final effect is operated under vacuum conditions. The temperatures and pressures throughout the entire system are determined by the vacuum applied, the fresh steam pressure, and the fresh steam flow rate relative to the flow of black liquor.

The fresh steam used in the first effect usually is extracted from turbines. The final effect normally operates at less than atmospheric pressure. In a six-effect system, the first effect may operate at about 30 psig and a steam temperature of 275° F. The final effect can be at a vacuum of 22 in. of mercury and a steam temperature of 150° F. The main advantage of a multiple-effect system is that it is possible to use the latent heat of vaporization in the inlet fresh steam several times over, and reduce overall steam requirements. The ratio of weight of water evaporated to weight of steam entering the system is called the steam economy or steam efficiency. For a six-effect system, a typical steam economy may be 5 to 5.5 lb of water evaporated per lb of fresh steam.

There are several different liquor flow configurations. A common configuration is to feed the weak black liquor simultaneously to the two final effects, which then operate in parallel, and then proceed countercurrent to the steam flow. As the liquor moves from one effect to another, the pressure increases, the boiling temperature increases, and the solids increase. Finally, the strong black liquor leaves the first effect. Flow strategies vary depending upon the evaporator system design, the desired percent solids in the product liquor, and the properties of the black liquor being processed. Properties of black liquor that are important (both in evaporation and in burning) are viscosity, boiling point elevation, density, heat capacity, surface tension, thermal conductivity, and scaling characteristics.

Build up of scale (also called fouling) is a major problem in evaporation. It can occur on either the liquor side or the steam side of the tubes. Scaling decreases the rate of heat transfer and requires shutting down the units for washing and cleaning. To prolong the interval between shutdowns, liquor flow to the different effects is sometimes switched. Lower solids liquor is sent to "wash" away the deposits left by a higher solids liquor. Scaling can also lead to an increase in the corrosion rate of the tubes.

Tall oil removal is required when evaporating liquors from the pulping of softwood species. This reduces scaling and other problems in the evaporators and recovers the commercial value of the tall oil. Tall oil is a mixture of resin and fatty acids from the wood. It is present as a soap in the black liquor. It floats when the black liquor solids content is in the range of 24% to 28% This separation is called "salting-out"; the soap is skimmed in a tank that follows the effect number 2, 3, or 4. The soap recovered is acid-treated to produce tall oil. When pulping hardwood species, soap skimming is usually unnecessary.

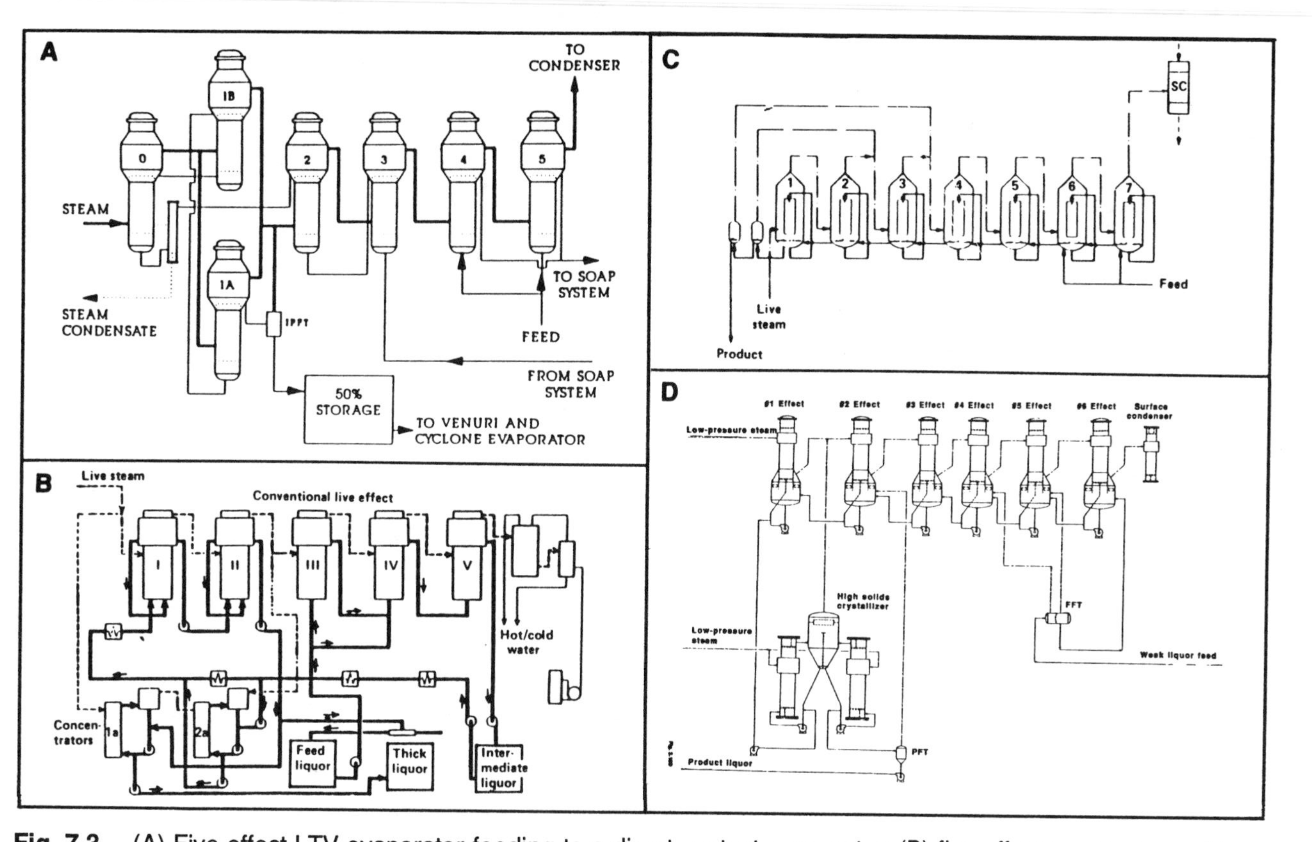

Fig. 7.3. (A) Five effect LTV evaporator feeding to a direct contact evaporator; (B) five effect evaporators with concentrators; (C) free flow falling film seven effect evaporator for full end concentration (Courtesy Rosenblad Corp.); (D) falling film evaporators with crystalizer concentrator. (Courtesy HPD, Inc.)

There are a number of evaporator designs available today (see Figure 7.3). The most common type has been a rising film evaporator, called the long tube vertical (LTV) evaporator (see Figure 3A). Liquor enters at the bottom. Due to the lower pressure at the top, the liquor rises up through the tubes where it picks up heat from the steam and begins to boil, forming a liquor-vapor mixture. The mixture overflows the tube section at the top, and is separated. The vapor continues to the next evaporator effect and the thickened liquor to another evaporator. The heating steam, or vapor, comes from the previous effect. It enters at the top of the unit where it releases heat to the liquor in the tubes and is condensed. The condensate is removed at the bottom of the tube section. Five- or six-effect LTV units are usually limited to 45% to 50% solids. Above that, the liquor becomes too viscous for upward flow. Also, severe scaling occurs due to reduced solubility of the sodium salts. Liquor from LTV systems are usually further concentrated in either direct contact evaporators (once common in older mills only) or in concentrators. The LTV liquor ends up in the range of 65% solids.

Direct contact evaporators were once commonly used to increase solids from LTV evaporator systems to the 62% to 65% level required for burning. Two types were used, a cascade type and a cyclone (venturi) type. Both used hot flue gases from the furnace in direct contact with the strong liquor. Flue gas temperatures are reduced from about 500° to 300° F. The flue gas will react with organic and inorganic sulfide compounds in the black liquor to form odorous gases called TRS (Total Reduced Sulfur - hydrogen sulfide and mercaptans). To avoid that, the black liquor was oxidized with air before evaporation in the LTV's. Black liquor oxidation (BLOX) formed sodium thiosulfate. The reaction is reversible and odors often persisted even after black liquor oxidation. BLOX is expensive to install and operate. DCE's (direct-contact evaporators) and BLOX units are being phased out in favor of falling film (FF) and concentrators.

Today, about 20% of all evaporators systems in North America are of the falling film type (FF), usually with six or seven effects. They deliver a product at 68% to 75% solids. These are either all FF types, or have one final effect with forced circulation (see Figures 3C and 3D). Falling film systems will increase as older LTV systems are retired. As the name suggests, the liquor falls downward as a film. This occurs either on the inside of tubes or the outside of plates. FF evaporators comprise a liquor distribution system, heating element, vapor body (optional) and separator system. A pump is usually provided to circulate a portion of the liquor from the bottom to the top. Circulation improves heat transfer control. FF evaporators have fewer scaling problems than LTV systems, especially at higher solids.

Special evaporator equipment may employ the forced circulation concept, such as the evaporator/crystallizer. They are being used to concentrate to 75% to 78% solids. Levels of 80% solids or higher are being produced in a few mills today.

Concentrators, operating with a feed liquor of 50% or more, are being used. These are merely evaporators employing live steam. Vapors are removed in a flash tank.

The Recovery Furnace (or Boiler)

Functions

The recovery boiler is the most expensive and most important unit in a pulp mill. It is often the reason for production limitations. Safe and consistent operation are essential for productivity and profitability.

The functions of the boiler are:

- Remove the water remaining in the strong black liquor
- Burn the organic matter in the black liquor
- Produce steam from the combustion gases
- Reduce the sulfur compounds to sulfides
- Recover the inorganic salts as a molten smelt.

These functions must be done while complying with strict environmental regulations regarding air emissions. Air emission regulations apply for SO_2 and for TRS (Total Reduced Sulfides) in the exit flue gas. Safety is another major consideration and is discussed later.

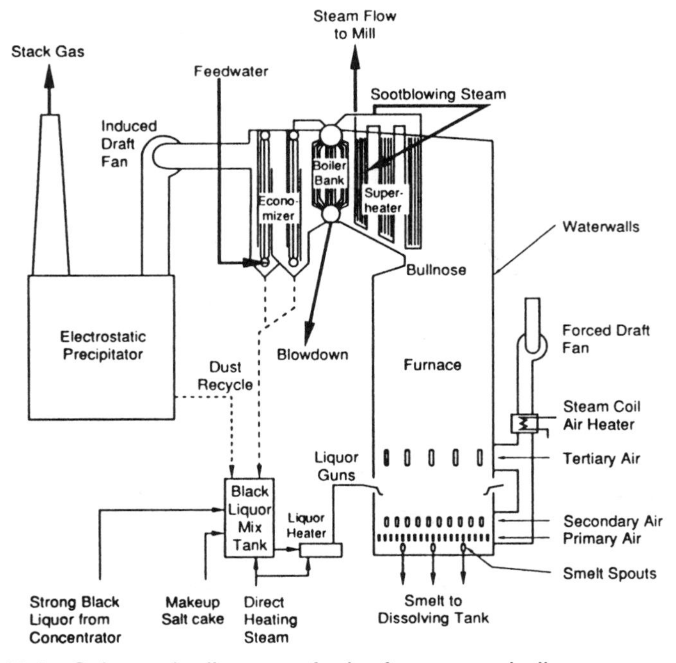

Figure 7.4 Schematic diagram of a kraft recovery boiler

Description of the Recovery Boiler

All modern recovery furnaces use a similar arrangement in which the processes of black liquor drying, pyrolysis, combustion of organics, and reduction of inorganic salts take place within a completely water-cooled enclosure (see Figure 7.4). Older recovery boiler designs had tangent tubes or tubes with flat studs backed by refractory; these constructions used thick casing plates on the outside in order to keep the flue gases within the furnace.

The modern furnace has a welded wall construction in which the space between the individual tubes is closed by a metal fin bar that is fully welded to both adjacent tubes. This "waterwall" usually has 2.5 in. OD tubes on 3 in. centers or 3 in. OD tubes on 4 in. centers. The construction is continuous from the furnace floor, up to and including the furnace roof. The resulting furnace is gas and smelt tight; a significant improvement over older designs in terms of safety and maintainability. Figure 7.5 illustrates various wall constructions.

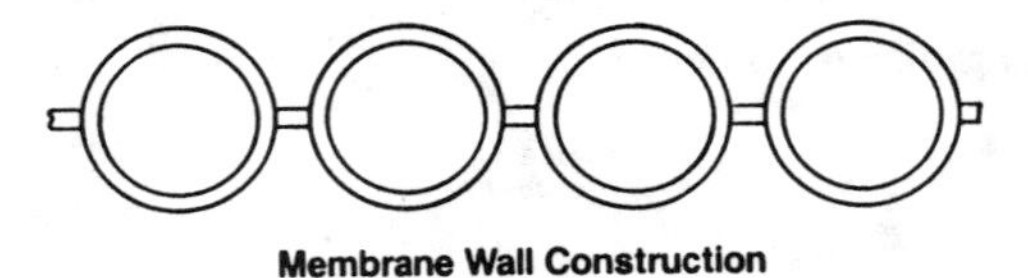

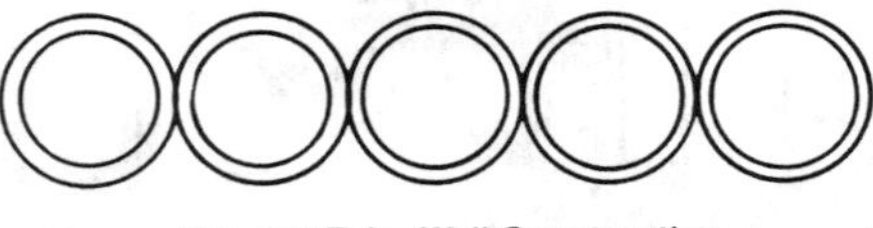

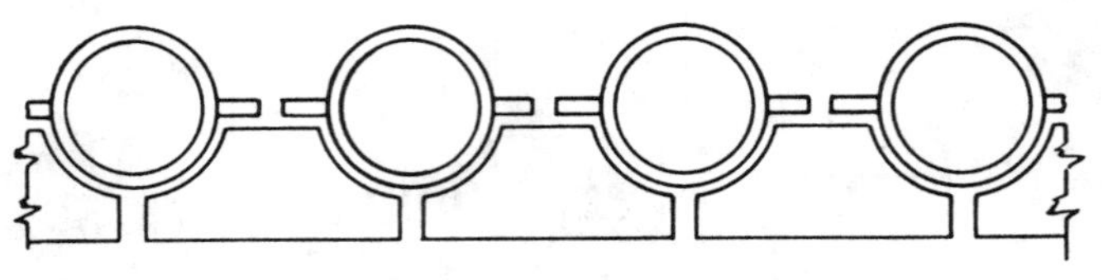

Figure 7.5 Various furnace wall construction

The entire furnace, complete with insulation and casing, is supported from the top by a steel or concrete structure. This accommodates the furnace's downward expansion of several in. when the furnace is heated up to operating pressures and temperatures.

Black liquor, auxiliary fuel for start-ups, and combustion air are admitted through a series of openings in the gastight waterwalls. These openings are usually created by offsetting one or more of the tubes and attaching an appropriate nozzle. Openings for smelt spouts, liquor firing guns, observation doors, access doors, and soot blowers are also made through offsetting tubes.

The waterwall tube material normally used is ordinary carbon steel. Below boiler water operating pressures of about 900 psig, it is not unusual to leave the plain carbon steel tubes bare, although the thickness may increase in the lower 30 ft or so of the furnace. Above about 900 psig, corrosion rates increase rapidly. Protection against this includes the use of pin studs, metallizing the tubes, weld overlaying, chromizing, or composite tubes. Composite or chromized tubing are used in most modern boilers as these do not involve attachments or coating. The initial cost is high, but maintenance costs are lower than the other alternatives. Figure 7.6 shows pin studs and composite tubing.

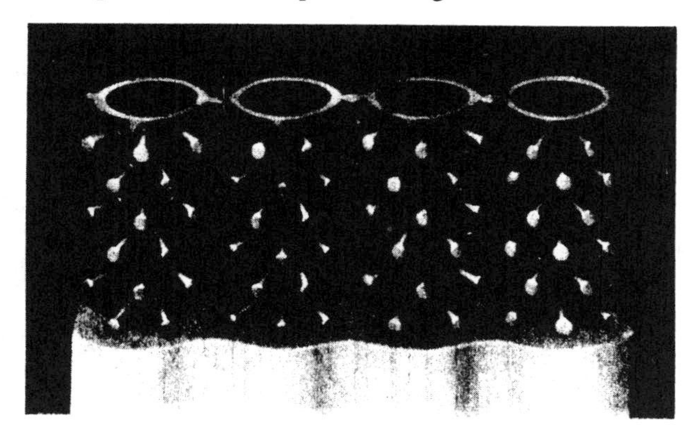

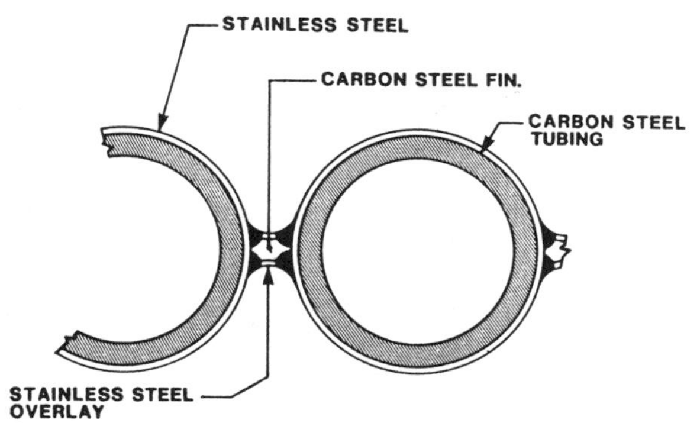

Figure 7.6

(Above): Pin studs protect against corrosion

(Below): Composite tubing demonstrates superior performance, lower maintenance.

Liquor combustion

The concentrated black liquor is pumped to a mix tank where make-up chemicals and recovered ash (called flyash) can be added. It passes through a heat exchanger to increase the temperature to about 270° F. This reduces the viscosity of the liquor, permitting it to be fired to the furnace through liquor nozzles called guns. Unlike fuel oil, which is sprayed in an atomized form, black liquor is sprayed in the form of coarse droplets. These droplets may fall to the hearth or be sprayed onto the walls of the furnace (spraying to the walls is somewhat obsolete today). In either case, the droplets dry in the presence of the hot combustion gases, partially burn and eventually fall onto the char bed. Liquor firing strategies vary depending upon the furnace manufacturer and the nature of the black liquor itself.

Preheated air is drawn into the boiler, usually at three locations, by forced draft fans. Primary and secondary air inlets are located below the liquor firing guns. The amount of air introduced here is slightly below the stoichiometric amount required for complete combustion. Reducing conditions in the char bed are essential to form sulfides from the oxidized sulfur compounds present (sulfates). Some of the primary reactions in the char bed are:

Formation of Na_2S

$$Na_2SO_4 + 2C = Na_2S + 2CO_2$$

$$Na_2SO_4 + 4C = Na_2S + 4CO$$

Formation of Na_2CO_3

$$Na_2O + CO_2 = Na_2CO_3$$

$$Na_2S + H_2O + CO_2 = NA_2CO_3 + H_2S$$

The hot combustion gases pass through the steam generation part of the boiler; the superheater, the boiler bank and the economizer. Assisted by a induced draft fan, the gas enters an electrostatic precipitator where entrained ash is removed. The ash is sent to the mix tank.

Air flows

As mentioned earlier, combustion air is preheated and enters the furnace at three locations. From the bottom up, these are the primary, secondary, and tertiary air inlets. It is important to maintain an accurate air supply. Two major operating variables are total air flow and the ratio of primary air to total air.

Primary (and secondary) air are introduced below the location of the liquor firing guns. These supply the required combustion air for the char bed and the lower portion of the furnace. Less than the stoichiometric amount of air needed for complete combustion is essential for the endothermic reaction to form sodium sulfide from the sodium sulfate. The tertiary air supplies the air needed to complete the combustion of the volatile gases emitted from the droplets of black liquor fed to the boiler and from the char bed itself. Insufficient air results in incomplete combustion, the formation of more odorous gases, and the danger of a "blackout" of the char bed as the bed literally suffocates.

Excessive air is thermally inefficient and leads to fouling of the heat transfer surfaces and plugging of gas passages due to excessive carryover of partially burnt liquor/smelt drops, and fine fume or dust particles. Entrainment of macroscopic liquor particles is generally considered to be more critical, but dust can also contribute to fouling if gas temperatures are high entering the closely spaced tubes at the front of the generating bank. Some carryover and dust is always present, so sootblowing is required to keep the boiler from plugging.

The boiler is under a balanced draft. Forced-draft fans supply the air to the furnace, and the flue gases are drawn through the generating banks with an induced draft fan.

Steam generation

Boiler feed water, heated to about 250° F, enters through the economizer bank to pick up heat from the flue gases. The water flows through the boiler bank. That can be either two drums with cross-flow, or a single drum with parallel flow. The temperature of the flue gas entering this section has to be below the critical temperature where ash stickies can deposit (about 1300° F if no potassium or chlorides are present; 1000° F if they are present). From there, the water flows through downcomers to a system of distribution headers below the furnace. The water then rises through the tubes in the waterwalls where some steam is formed. The steam and water are separated with the steam going to the superheaters. The superheated steam leaves the recovery boiler at 1250 psig/900°F or higher. This steam produces electrical energy in the turbines and is a source for the mill's steam requirements.

Smelt dissolving

Molten smelt leaves the furnace at about 1500° F through a number of water-cooled spouts. Steam is commonly used to shatter the smelt into small particles before it enters the dissolving (smelt) tank. Weak wash water from the causticizing operation is the main supply of water. Green liquor is formed at this point. Control of the flow, temperature and concentration of the sodium carbonate and sodium sulfide in the liquor is important to the causticizing operation. Some mills provide a second tank, called a surge tank or a stabilization tank, to even out fluctuations.

The interaction of molten smelt with water is a violent one. It is estimated that one lb of water vaporizing in one millisecond releases the equivalent energy of ½ lb of TNT. Smelt-water explosions usually involve 5 to 25 lb of water and are the main cause of disastrous explosions in recovery boiler operations.

Safety

Other causes of explosions in recovery boilers are water from leaky tubes in the waterwalls contacting the char bed, and from auxiliary fuel usage on start-ups.

Decades ago, the industry formed a Black Liquor Recovery Boiler Advisory Committee (BLRBAC, or Blarback) to investigate causes of disastrous explosions and to recommend preventative actions. Blarback is an international committee composed of representatives of paper companies, boiler manufacturers, insurance companies, and professionals from various worldwide institutions. Today, all boilers have detailed emergency shutdown procedures.

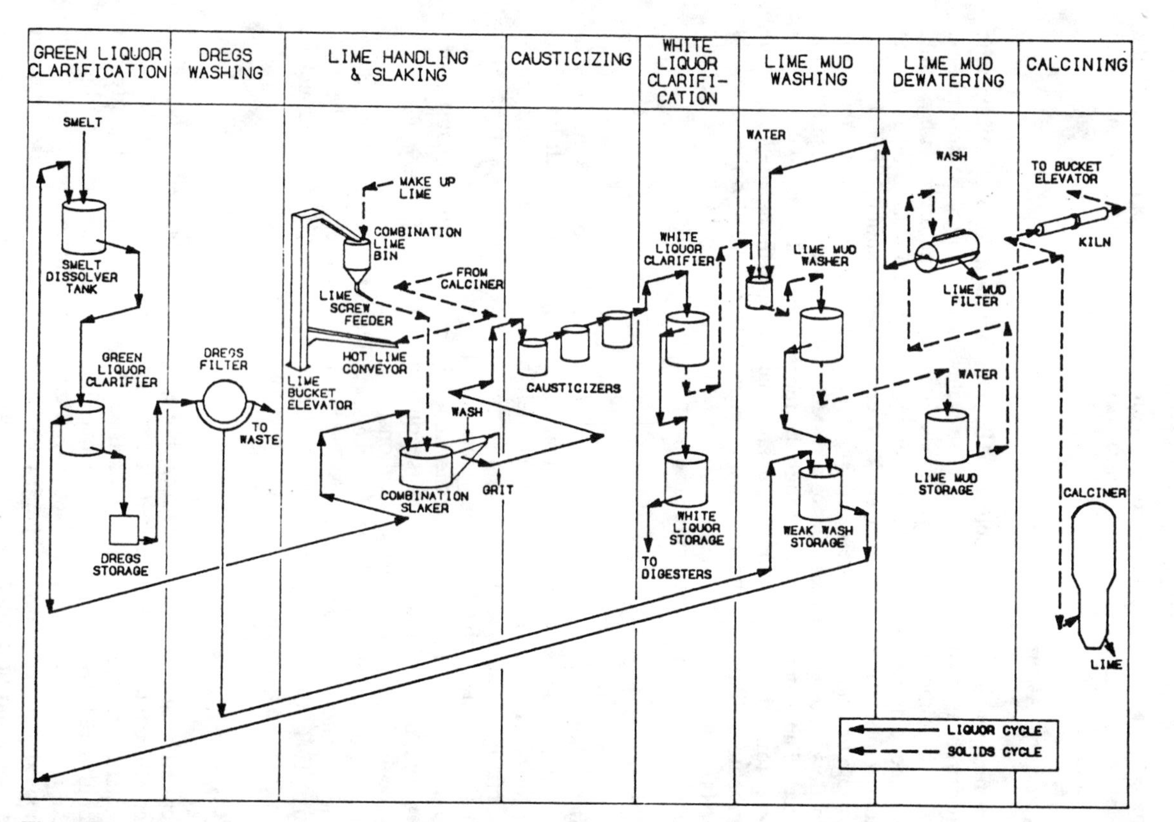

Fig. 7.7 Unit operations flowsheet for sedimentation type kraft recausticizing system

While explosions have become a relative rarity today, danger remains whenever smelt and water contact each other in an uncontrolled situation.

Causticizing (Also called Recausticizing)

The conversion of sodium carbonate into sodium hydroxide is one of the oldest chemical processes. It has been done on a continuous basis for more than 75 years. While the chemistry remains unchanged, there have been continuous developments in the equipment used and the control strategies employed. Variations in the operation of any one step will influence each subsequent step.

The objectives of the process are to:

- Convert sodium carbonate to sodium hydroxide
- Recycle the lime "mud" (calcium carbonate) formed into reusable lime (calcium oxide)
- Remove the solid impurities from the green and white liquors.

Figure 7.7 illustrates a typical kraft recausticizing system.

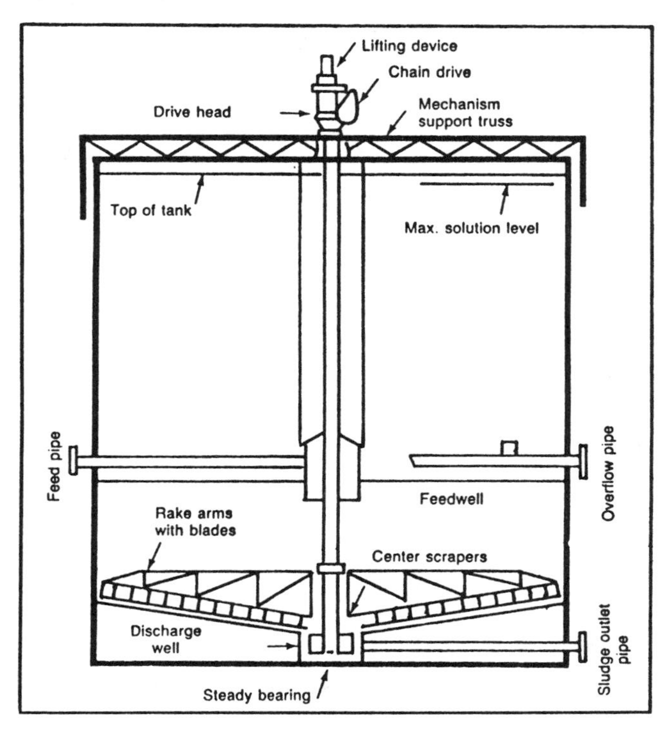

Figure 7.8 Green liquor clarifier with storage

Chemistry

The reactions occurring in the causticizing process are:

- $Ca(OH)_2 + Na_2CO_3 = CaCO_3 + 2NaOH$ (causticizing)
- $CaCO_3 + Heat = CaO + CO_2$ (lime reburning)
- $CaO + H_2O = Ca(OH)_2$ (slaking)

All of the calcium compounds involved are solids. Much of the equipment used in causticizing is involved with the separation of these solids from the liquors. The Na_2S in the green liquor does not react during causticizing.

Green liquor clarification

Green liquor from the smelt tank (or the stabilization tank) contains particles such as unburned carbon, solids from the furnace construction, and perhaps some calcium compounds. These are called dregs. Typically, about 1,000 ppm of dregs are present. Their removal is important in subsequent operations and for the purity of the white liquor formed. This was usually done in a green liquor clarifier (see Figure 7.8). The dregs settle and the clear green liquor is decanted from the top. The dregs slurry is pumped from the bottom, mixed with wash water, settled, and filtered in order to thicken them. The thickened dregs are normally sent to a landfill. The wash water is returned to the smelt tank.

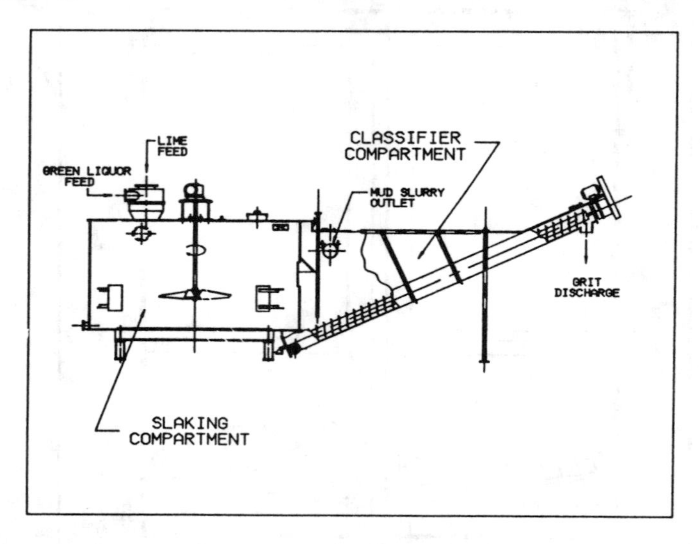

Figure 7.9 Slaker classifier illustrating wetted-wall lime inlet and screw grit removal technique.

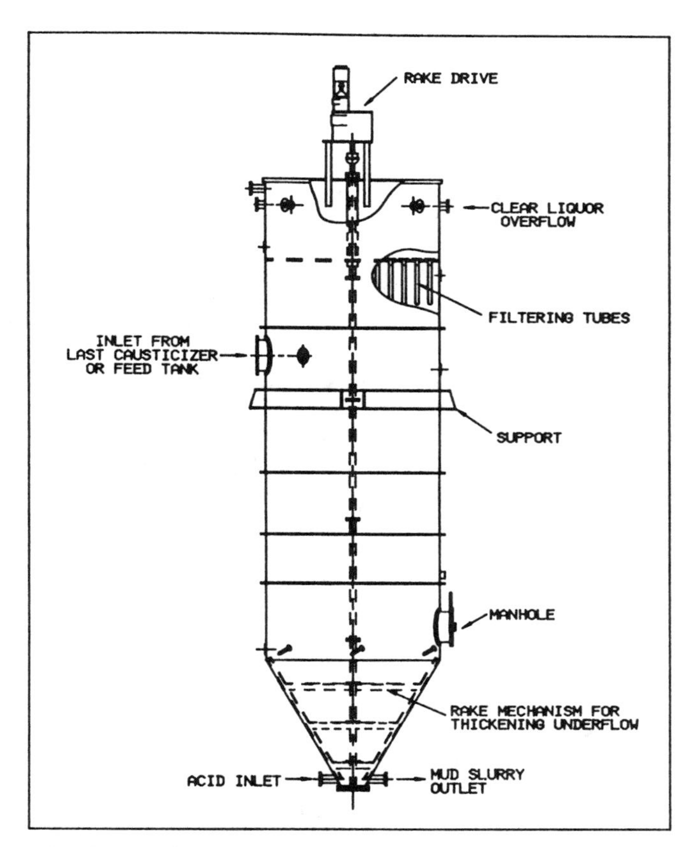

Figure 7.10 Pressurized tubular filter design

Slaking and causticizing

The clarified liquor is pumped to a "slaker" where it is mixed with reburnt lime from the lime kiln plus the required amount of fresh (make-up) lime (see Figure 7.9). The slaking reaction occurs here. The slaker has two functions: slaking the lime and removing impurities found in the lime used. These impurities are called grits and are removed, washed, and sent to landfill.

While the causticizing reaction begins in the slaker, it is completed in a series (usually 3) of agitated tanks called causticizers. Their total retention time is in the order of 90 min. The causticized liquor is now called white liquor. It still contains all of the lime "mud" (calcium carbonate) formed.

White liquor clarification

White liquor clarification (removal of the mud) was normally done in equipment essentially the same as for green liquor clarification as shown in Figure 7.8. Both green liquor and white liquor clarifiers normally have sufficient height to serve as storage tanks as well. The lime mud, at about 35% to 45% solids, is removed by a slow speed rake.

Pressure filters

Pressure filters are increasingly being used to replace clarifiers, both for green liquors and white liquors. They require less space and are said to deliver a purer product liquor. Figure 7.10 shows a typical pressure filter. This uses polypropylene felt type filter media over perforated stainless steel tubes.

Filtration occurs on the tubes in the upper part of the feed section; the portion below serves for mud storage and thickening. The operation is not continuous as filtration time is interrupted for backwashing the tubes. A typical cycle is 5-10 min for filtration, 5 seconds for back-flushing and 30 seconds for settling of the discharged solids. External valves are used for regulation.

The polypropylene filter media requires periodic cleaning with an acidic wash, usually every 4-16 weeks.

Mud washing and thickening

The mud from the clarifier or from the pressure filters is washed to remove the soluble cooking chemicals and thickened for burning in the lime kiln. A variety of equipment is available. Precoat vacuum drum filters were commonly used in the past. Pressure filters and disc filters are now being used as well. The object is to produce 60% to 75% solids mud for the kiln from the 40% or so solids exiting the white liquor clarifier.

Lime kiln

Lime reburning is done almost exclusively in kilns. They are more energy efficient than fluidized bed calciners. Modern lime kilns require about six million BTU's per ton of lime product, about 20% less than a fluidized bed reactor.

The rotary kiln is a long, cylindrical refractory-lined vessel, sloped slightly from the feed end to the discharge end, rotating at about one rpm (see Figure 7.11). The four stages in lime reburning are shown in Figure 7.12.

In the drying stage, water is evaporated from the wet mud, which usually enters at a solids content of about 67% to 70%. Drying is accomplished as the mud passes a series of metal chains attached to the kiln shell. The chains are designed to lift and to dry the mud to about 95% to 100% solids. The chains also act as a dust curtain to minimize lime dust leaving with the flue gases.

In the second stage, the dried mud is heated by heat transfer devices such as tumblers and lifters that are attached to the kiln shell. These lifters and tumblers

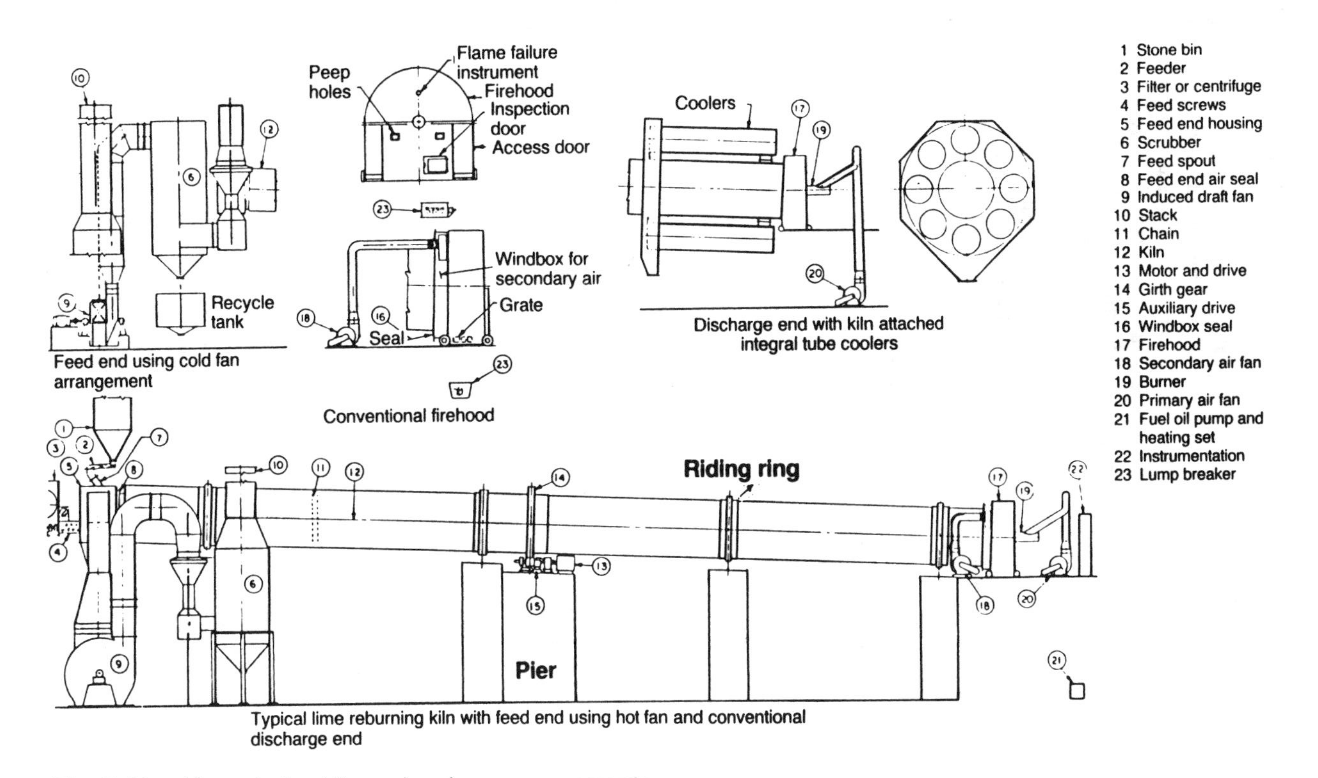

Fig. 7.11 Lime sludge kiln and various arrangements

stir the lime and promote mixing of the material with the hot gases. Some kilns use mixing shields and steel bars in this zone to further improve heat transfer.

Calcination occurs in the third zone, where carbon dioxide gas is released and calcium oxide pellets begin to form. The ideal lime pellet is about ¾ in. in diameter. If the mud is not properly dried, large lime balls can form. Excess sodium entering with the mud aggravates the problem of large balls. The calcination reaction is temperature dependent; a minimum of 1500° F is required, while temperatures over 2100° F result in overburnt lime and poor slaking.

Cooling, the final stage, occurs as the pellets pass under the burner and move to the discharge end of the kiln. The hot lime pellets will cool down, exchanging heat with the incoming secondary air flow. Modern kilns employ tube coolers where the secondary air enters, cooling the exit lime to about 350° F to 400° F.

The flue gas will contain dust particles and perhaps some SO_2 and TRS gases. Regulations require their removal. Wet scrubbers, employing water, were used for scrubbing these gases. Today, electrostatic precipitators, venturi type wet scrubbers, and cyclone separators (used in combination with a scrubber) are commonly used.

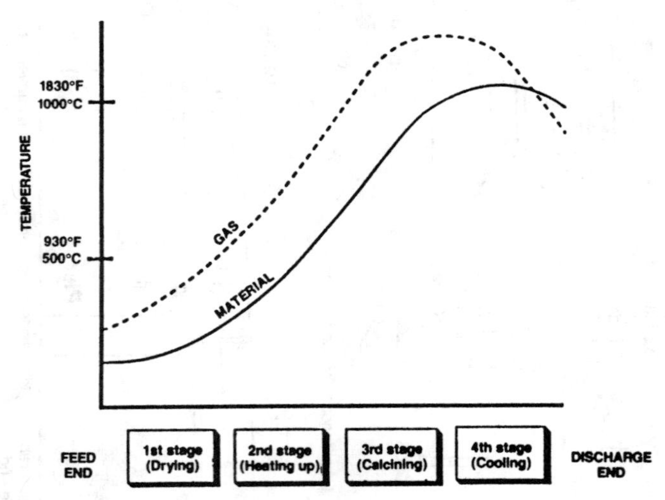

Figure 7.12 Kiln temperature profile
Courtesy Termorak, Inc.

QUESTIONS

1. Specify the functions of the kraft recovery process.
2. List the five main steps of the kraft recovery process.
3. Discuss the content of a typical mill black liquor.
4. Define the objective of black liquor evaporation.
5. Describe the operation of a multiple-effect evaporator system.
6. Name two different evaporator designs and explain how they function.
7. Discuss the construction of a recovery boiler.
8. Describe the process of liquor combustion.
9. Describe how steam is generated in a recovery boiler.
10. Identify some possible problems related to flue gas flow and how they can be solved.
11. Identify the condition causing the majority of furnace explosions in the past.
12. Describe the purpose of the causticizing step.
13. List the major steps in recausticizing.
14. Describe how green liquor is formed.
15. Describe the slaking and causticizing operation.
16. Write the reactions for slaking and causticizing.
17. State the purpose of green liquor and white liquor clarification.
18. State the reason for lime burning.
19. List the four steps in lime reburning.
20. Discuss the design of a lime kiln.

Appendix

Fundamental concepts

Atoms

Atoms are the smallest building blocks in chemistry. One atom consists of a positively charged nucleus and negatively charged electrons that circle around the nucleus.

The nucleus consists of positively charged protons and neutrons without any charge. There is no way one can chemically break down an atom and separate its components. An atom always contains an equal number of protons and electrons and thus has no external charge.

A substance that contains only one kind of atom is called an *element.* There are about 100 elements existing in nature, for example, iron, oxygen, carbon, and sulfur. Each element has its own letter symbol.

The difference between different kinds of atoms is in how many protons, neutrons, and electrons they contain. The more protons and neutrons an atom's nucleus contains, the heavier it is; that is, it has a higher *atomic weight.*

A hydrogen atom has an atomic weight of 1.0u since its nucleus consists of one proton only. A carbon atom has an atomic weight of 12.0u since its nucleus consists of six protons and six neutrons. 1.0u equals $1.66 \cdot 10^{-27}$ kg, which is an extremely small number.

Molecules

Atoms can react with each other and form *molecules.* Substances made of molecules are called *compounds.* Water and salt are examples of compounds.

Atoms react with each other because some atoms have a tendency to attract other atoms' outer electrons, and some have a tendency to give away their outer electrons to other atoms. Molecules can also react with each other to form separate, larger molecules.

Chemical bonding

There are two different kinds of bonds that can be formed in a chemical reaction.

The *covalent bond* is the strongest bond. It is formed between atoms that can attract each other's outer electrons. The two atoms end up sharing two electrons, each having contributed one. These two shared electrons hold the atoms together with strong electric forces.

An example of covalent bonding is when two identical atoms react with each other, for example, when two oxygen atoms react to form oxygen gas:

$$2\ O \rightarrow O_2$$

Water is another example. One oxygen atom is bonded to two hydrogen atoms with two covalent bonds:

H_2O.

The other kind of bond that can be formed in a chemical reaction is the *ionic bond.* It is formed when one atom strongly attracts the other atom's outer electrons and the other atom has a strong tendency to give away its outer electrons.

Here the electrons actually are exchanged, and two electrically charged *ions* are formed. These two ions are attracted to each other because of their opposite charges. This attraction is what creates an ionic bond. An example of ionic bonding is the joining of sodium and chlorine to form table salt:

$$Na + Cl \rightarrow NaCl$$
$$NaCl + H_2O \rightarrow Na^+ + Cl^-$$

This ionic bond is so weak that when salt is added to water, it dissolves and the water molecules interact with the ions. One can never separate positive and negative ions in solid form, though.

Another example of ionic bonding is the formation of sodium hydroxide (soda) from sodium and water:

$$H_2O + Na \rightarrow Na^+OH^- + 1/2\ H_2$$

The unreacted hydrogen leaves as hydrogen gas.

Molecules can also form positive and negative ions. As a matter of fact, water reacts with itself to form H_3O^+ and OH^- at all times:

$$2\ H_2O \rightarrow OH^- + H_3O^+$$

Ions

Certain atoms will readily give up or accept electrons to form an electrically charged atom or *ion.* For example, sodium (Na) will readily lose one electron to form positively charged sodium ions (Na^+). The reaction is:

$$Na \rightarrow Na^+ + \text{one electron } (e^-)$$

This tendency of sodium (and some other closely related metals) to give up one electron is so great that when sodium metal is placed in water, a violent reaction occurs.

Likewise, elements like chlorine (Cl) will readily accept electrons (e^-) to form a negatively charged chlorine ion. The reaction is:

$$Cl + e^- \rightarrow Cl^-$$

A solid substance consisting of ions, like sodium chloride, is called an ionic solid. When ionic solids are dissolved in water, they separate into individual ions:

$$NaCl \xrightarrow{(water)} Na^+ + Cl^-$$

Some elements (mostly metals) can form more than one type of ion. For example, iron (Fe) can form ions having either a +2 or a +3 charge.

Certain groups of atoms will collectively act as a single ion. Some examples are:

SO_4^{2-}	sulfate ion
CO_3^{2-}	carbonate ion
OH^-	hydroxyl ion

In chemical reactions, these ions generally remain intact and thus behave as ions consisting of a single atom. An example to illustrate this is the reaction between slaked lime ($Ca(OH)_2$) and sodium carbonate (Na_2CO_3).

Na_2CO_3	+	$Ca(OH)_2$	$\longrightarrow$	$2\ NaOH$	+	$CaCO_3$
sodium carbonate		calcium hydroxide		sodium hydroxide		calcium carbonate

Nomenclature

As mentioned earlier, each element has its own letter symbol. It is often the one or two first letters of its (Latin) name. Here is a list of some of the more common elements that are involved in kraft pulping chemistry.

Symbol	Name
H	hydrogen
O	oxygen
C	carbon
S	sulfur
Na	sodium
Ca	calcium
Au	gold
Ag	silver
Cl	chlorine

Molecules and ions also have letter symbols. One simply combines the symbols of the atoms of which the molecules consist. By knowing that the symbol for water is H_2O, one can tell that water consists of two hydrogen atoms and one oxygen atom. The following is a list of some molecules and ions common in nature and kraft pulping.

Symbol	Name	Symbol	Name
H_2O	water	Cl^-	chloride ion
$NaCl$	sodium chloride (table salt)	OH^-	hydroxide ion
$NaOH$	sodium hydroxide (caustic soda)	SH^-	hydrosulfide ion
Na_2CO_3	sodium carbonate		
Na_2S	sodium sulfide	S^{2-}	sulfide ion
H_2S	hydrogen sulfide		
$Ca(OH)_2$	calcium hydroxide (slaked lime)	SO_4^{2-}	sulfate ion
CaO	calcium oxide (quicklime)	CO_3^{2-}	carbonate ion
$CaCO_3$	calcium carbonate		
HCl	hydrochloric acid		
C_2H_5OH	ethanol (alcohol)		
$(CH_3)_2CO$	acetone		

The letter symbols are used when one wants to write down chemical reactions. As one can see by the following comparison, it is a convenient way to convey information:

$$2\ H_2 + O_2 \rightarrow 2\ H_2O$$

Two molecules of hydrogen gas react with one molecule of oxygen gas to form two molecules of water.

The molar concept

Atoms and molecules are so small that it would be difficult to make calculations based on their weight. Therefore, the concept of the mole was introduced. One mole of a substance always contains $6.023 \cdot 10^{23}$ units. For example, one mole of oxygen gas contains $6.023 \cdot 10^{23}$ oxygen molecules.

The basic definition of one mole is the number of atoms a 12.0 g sample of carbon contains. Carbon has an atomic weight of 12.0u. One can therefore say that one mole of carbon weighs as much in grams as does one single atom of carbon expressed in atomic weight.

The same relationship exists for all substances. For example, one mole of water, having a molecular weight of 18.0u weights 18.0 g.

The number of atoms or molecules in one mole, $6.023 \cdot 10^{23}$, is a very large number. If one were to calculate the age of the earth in seconds and multiply this number by four million, the answer would be approximately as large as the number of atoms or molecules in one mole.

Molar weights

The molar weight of an element or compound is simply the weight in grams of one mole of the substance.

The *molar weight* is used to calculate concentrations of solutions or to calculate how much of two substances are needed for a certain reaction.

The following is a table of the molar weights of some common elements and compounds: (Grams per mole is abbreviated as g/mole.)

Name	Symbol	g/mole
H	hydrogen	1.0 g
H_2	hydrogen gas	2.0 g
O	oxygen	16.0 g
O_2	oxygen gas	32.0 g
H_2O	water	18.0 g
C	carbon	12.0 g
S	sulphur	32.1 g
Na	sodium	23.0 g
Ca	calcium	40.1 g

Notice that the molar weight for oxygen gas O_2 is exactly double the molar weight of oxygen atoms, O. This is because one oxygen molecule weighs as much as two oxygen atoms.

Example of calculations using molar weights

Problem: You have 24 g of carbon you want to burn to form carbon dioxide, CO_2. How many grams of oxygen gas will be consumed?

First write down the reaction: $C + O_2 \rightarrow CO_2$

This can be read as "one mole of carbon reacts with one mole of oxygen gas to form one mole of carbon dioxide." (Since one mole always contains the same number of units.)

How many moles of carbon do you have?

$$\frac{24\text{ g}}{12\text{ g/mole}} = 2\text{ moles}$$

We have two moles of carbon and thus need two moles of oxygen. How many grams of oxygen is that?

$$2 \text{ moles} \times 32 \text{ g/mole} = 72 \text{ g oxygen gas}$$

We have now calculated that we need at least 72 g oxygen gas to fully burn 24 g of carbon to form CO_2.

Solutions

A solution results when one substance, the *solute,* dissolves in a *solvent.* For example, when sugar is added to water, it disappears. This is because the sugar crystals are broken down to individual molecules and the crystals dissolve. The sugar, in this case, is the solute, and the water is the solvent. The resulting mixture is a solution. The individual sugar molecules become totally and uniformly dispersed among the water molecules.

The amount of solute dissolved in a solvent is known as the *concentration.* Concentration can be expressed in many different ways:

a. *Weight of solute per volume of solution.*
Example: lb/gal or g/L

b. *Percentage composition*
Defined as the weight of solute per 100 weight units of solution.
Example: 100 lb of a 10% sugar solution contains 10 lb sugar.

c. *Molarity*
Molarity is defined as the number of moles of solute per liter of solution.
Molarity is abbreviated as "M."
Example: One liter of 2.0 M hydrochloric acid solution contains two moles of hydrochloric acid.

pH

pH is a measure of how acidic or alkaline a solution is. What really determines pH is the balance between hydrogen ions (H^+) and hydroxide ions (OH^-) in the solution.

A solution with an excess of hydrogen ions is acidic while a solution with an excess of hydroxide ions is alkaline. If a solution contains equal amounts of hydrogen ions and hydroxide ions, it is neutral.

Pure water is considered neutral, having a pH of 7. Solutions with a pH lower than 7 are acidic, with acidity increasing as pH is lowered. Solutions with a pH higher than 7 are alkaline, with alkalinity increasing as pH rises.

Each pH unit corresponds to a change in concentration of hydrogen and hydroxide ions ten times. Therefore, a solution of pH 2.0 is ten times more acidic than a solution of pH 3.0 and a kraft pulping liquor of pH 13 is 1,000,000 × (one million times) more alkaline than pure water.

Acids are compounds that release hydrogen ions when dissolved in water. Here is how hydrochloric acid reacts:

$$HCl \rightarrow H^{+} + Cl^{-}$$

Acids taste sour.

Bases are compounds that release hydroxide ions when dissolved in water. Here is how caustic soda reacts:

$$NaOH \rightarrow Na^{+} + OH^{-}$$

Strong solutions of a base have a slimy feeling. Strong bases and acids are very reactive and, therefore, dangerous. Avoid direct skin contact and always wear safety glasses when handling them.

Literature Sources

A: M.J. Kocurek and C.F.B. Stevens, Ed. *Pulp and Paper Manufacture, Volume 1: Properties of Fibrous Raw Materials and their Preparation for Pulping,* 3rd ed., Joint Textbook Committee of the Paper Industry, 1983.
B: J.V. Hatton, Ed. *Pulp and Paper Technology Series, No. 5: Chip Quality Monograph,* Joint Textbook Committee of the Paper Industry, 1979.
C: *Notes from Chip Preparation and Quality Seminar,* TAPPI, Washington, D.C. 1987.
D: J.P. Casey, ED. *Pulp and Paper Chemistry and Chemical Technology, Volume 1,* 3rd ed., John Wiley & Sons, New York, 1980.
E. D.W. Clayton, Ed. (?) *"Chemistry of Alkaline Pulping", Part I of Joint Textbook, Vol. 5; Kraft Pulping* (final draft).
F: *Sulfatmassetillverkning,* Sveriges Skogsindustriförbund, Sweden, 1986.
G: G.A. Smook, *Handbook for Pulp & Paper Technologists,* Joint Textbook Committee of the Paper Industry, 1982.
H: *"Alkaline Digester Systems", Joint Textbook, Vol. 5: Kraft Pulping* (final draft).
I: J.K. Perkins, *"Post Digester Treatment of Sulphate Pulp", Joint Textbook, Vol. 5: Kraft Pulping* (final draft).
J: J.K. Perkins, *"Washing, Screening, Cleaning, and Handling of Pulp", Joint Textbook, Vol. 5: Kraft Pulping* (final draft).
K: B. Cowan, *"Screening", Joint Textbook, Vol. 5: Kraft Pulping* (final draft).
L: T. Hooper and S.W. Hooper, *"The Screening of Chemical Pulp", Joint Textbook, Vol. 5: Kraft Pulping* (final draft).
M: B. Wikdahl, *"Centrifugal Cleaning", Joint Textbook, Vol. 5: Kraft Pulping* (final draft).
N: B. Cowan *"Pulp Handling", Joint Textbook, Vol. 5: Kraft Pulping* (final draft).
O: R.H. Crotogino, N.A. Poirier and D.T. Trinh *Tappi Journal,* 70(6): 95–103 (1987).
P: J.K. Perkins (ed.) *Brown Stock Washing Using Rotary Filters* Tappi Press, Atlanta, Ga., 1983..
Q: E.. Sjöström, *Wood Chemistry,* Academic Press, New York, 1981.